Buckwheat: Forgotten Crop for the Future

Buckwheat: Forgotten Crop for the Future

Issues and Challenges

Tanveer Bilal Pirzadah and Reiaz Ul Rehman

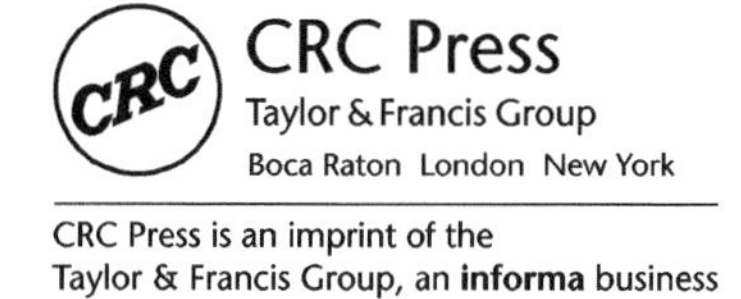

First edition published 2021
by CRC Press
6000 Broken Sound Parkway NW, Suite 300, Boca Raton, FL 33487-2742

and by CRC Press
2 Park Square, Milton Park, Abingdon, Oxon, OX14 4RN

CRC Press is an imprint of Taylor & Francis Group, LLC

Library of Congress Cataloging-in-Publication Data

Names: Pirzadah,Tanveer Bilal, author. | Rehman, Reiaz Ul, author.
Title: Buckwheat forgotten crop for the future : issues and challenges / Tanveer Bilal Pirzadah, Reiaz Ul Rehman.
Description: First edition. | Boca Raton, FL : CRC Press, 2021. | Includes bibliographical references and index.
Identifiers: LCCN 2021005814 | ISBN 9780367543884 (hardback) | ISBN 9780367543938 (paperback) | ISBN 9781003089100 (ebook)
Subjects: LCSH: Buckwheat.
Classification: LCC SB191.B9 P57 2021 | DDC 633.1/2--dc23
LC record available at https://lccn.loc.gov/2021005814

ISBN: 9780367543884 (hbk)
ISBN: 9780367539388 (pbk)
ISBN: 9781003089100 (ebk)

Typeset in Times
by Deanta Global Publishing Services, Chennai, India

Dedication

This book is dedicated with much love and affection to my beloved parents

Contents

Foreword

Buckwheat is expected to be an even more important plant in agriculture and an ingredient in the cuisine and dietary habits of India and many other nations, including those of Asia, Europe and America. Besides the agricultural, cultural and culinary value of buckwheat, recently more emphasis has been put on its health-related and nutritional value: the well-balanced amino acid composition of its proteins, its dietary fiber, its mineral elements and vitamins, as well as its rich content of diverse antioxidants, mainly the flavonoids rutin and quercetin. Novel buckwheat dishes and products, including diverse pasta products, grain-based products, sprouts and young plants have potential in India and elsewhere. One of the reasons for buckwheat's potential is its relative resistance to pests and diseases and its ability to repress weeds. Buckwheat is thus suitable for biological (organic, ecological) cultivation. It does not contain gluten proteins, so it is safe for people who require a gluten-free diet. Buckwheat can increase food diversity; it evokes tradition and in a way revives the heritage of "the good old days." Two types of buckwheat are mainly used around the world: common buckwheat (*Fagopyrum esculentum*) and Tartary buckwheat (*Fagopyrum tataricum*). Which buckwheat species is used depends on the production zone and the method of utilization. Generally, in Europe, the United States, Canada, Brazil, South Africa and Australia common buckwheat prevails. This also holds true for most Asian buckwheat growing countries, e.g. Japan, Korea and the central and northern parts of China. Tartary buckwheat is grown and used in the mountain regions of the Himalayas. In northern India, Bhutan and Nepal, both types are known, although Tartary buckwheat is grown in harsher climatic conditions. Tartary buckwheat flour is recently increasingly used in preparing dishes due to the fact that it has an even higher rutin content than common buckwheat. Common buckwheat and Tartary buckwheat are used in different parts of the world to make various food products. Buckwheat groats are a prebiotic food because they can, for example, increase the lactic acid bacteria in the intestine due to its resistant starch content.

In this book, the authors summarize the origin, history and distribution of buckwheat throughout the world. They also discuss how buckwheat was introduced into the Kashmir valley from its place of origin via the old silk road. Further, they discuss the limitations and potential benefits of the cultivation of this ancient crop. The authors emphasize the rewards of reviving buckwheat culture, cultivation and utilization, based on its rich tradition and current knowledge about the importance and quality of buckwheat, including for the functional food sector. They also discuss strategies for buckwheat crop improvement using modern knowledge and approaches. From an agronomical point of view, buckwheat represents a good opportunity for eco-friendly cultivation, allowing a reduced use of conventional fertilizers and pesticides and without requiring transgenic genetic manipulation, to maintain ecological crops, widely acceptable for consumers around the world.

Ivan Kreft (D. Sc)
Professor of Genetics
University of Ljubljana, Slovenia
and
President
International Buckwheat Research Association (IBRA)

Preface

The globalization of agriculture and consequently its industrialization seems unrelenting, with adverse side effects felt throughout the world. These effects include, but are not limited to, technological development biased towards the use of only some high energy–demanding plant species and monoculture practices, and in this way the genetic diversity of the agriculture sector has been diminished. As a result, world food security has become largely dependent on only a handful of crops. Even if humankind has, over time, cultivated more than 10,000 edible species, nowadays only some 150 plant species are commercialized, 12 of which provide approximately 80% of the requirement for dietary energy from plant bio-resources; over 60% of the global requirement for protein and calories are met by just four species: rice, wheat, maize and potatoes (FAO 2005). Therefore the reduction in the number of crops upon which global food security and economic growth depend has placed the future supply of food and rural incomes at risk. Moreover, the ever-increasing population of the world is projected to reach 10.9 billion by 2050. Today, in a world of 7 billion people, agriculture is facing great challenges to ensure a sufficient food supply, while at the same time maintaining high productivity and quality standards. The narrow agricultural portfolio raises serious questions about how effectively major crops alone can contribute towards food security and poverty alleviation. The above-mentioned facts have profound environmental consequences and create concern for loss of crop varieties, provoking organizations and research scientists worldwide to retrieve, research and disseminate knowledge about the production and utilization of neglected, under-exploited and new plant species or so-called alternative crops. Addressing this concern requires broadening the focus of research and development to include a much wider range of crop species. Such crops must also have the potential to achieve a high value on the market due to the characteristics of their products. In this context, the International Centre for Underutilized Crops (ICUC) has identified buckwheat as having strong potential for domestication. Buckwheat (*Fagopyrum* spp.) is an ancient crop, which has long been grown in East Asia and the Himalayan region. It is a major staple food crop in high altitude zones, viz. Himalayan regions and Daliang Mountain in Southwest China. It is the most important crop grown in mountain regions above 1800m elevation, both for its grain and greens. Unlike common cereals, which are deficient in lysine, buckwheat protein is of excellent quality in terms of its essential amino acid composition. The upsurge in interest in buckwheat is based on its high nutritional profile qualities, including the high protein content of its grains, its flavonoid content, its ability to grow in marginal areas and its suitability for cultivation as an organic (biological or ecological) traditional crop. Buckwheat has been designated an important crop by UNESCO due to its dwindling cultivation and exploitation in the wild. Currently, buckwheat has gained worldwide importance in the functional food sector due to the presence of some its important bioactive constituents, such as phenolic compounds, phytosterols, fagopyrins, amino acids, lysine (particularly deficient in cereals), dietary fiber, lignans, vitamins, minerals, antioxidants, unsaturated fatty acids and other essential components like fagopyritols, which have immense potential for glycemic control in type II diabetics.

Since agricultural land is a limiting factor, one of the choices for buckwheat cultivation remains those marginal lands which are more or less infested with abiotic stressors. Anthropogenic activities have also caused soil deterioration, which leads to abiotic stress. This abiotic stress, which can include drought, temperature extremes, waterlogging, salinity, heavy metal toxicity and nutritive deficiency, directly and indirectly alters plant physiology, affecting plant growth and productivity and therefore causing economic losses in agriculture worldwide. Consequently, underdeveloped or developing countries find it extremely difficult to maintain countrywide food protection. Furthermore, climate change may have an effect on all aspects of food protection, as well as on food availability, accessibility, usage and stability. Thus, for food production, agro-systems must be adapted to climate change to ensure food security and stability. There is therefore a need to adopt recent biotechnological interventions for crop improvement and the domestication of buckwheat as an underutilized crop. We thus felt it was very important to publish a book on this important underutilized crop. This book tries to connect and catch up with modern developments in buckwheat research over time so as to fall in line with global trends. Limited information describing basic aspects of buckwheat's genetics, physiology, agronomy etc. has long been a hindrance to its development and promotion. Enhancement of the extent of its incorporation into the agricultural portfolio through appropriate species-specific interventions targeted at problems that hinder its domestication would be an important step towards its domestication and towards greater agricultural diversification.

This book is unique and would be an ideal source of scientific information for postgraduate students, research scholars, faculty and scientists involved in agriculture, the plant sciences, molecular biology, biochemistry, biotechnology and food technology.

We extend our appreciation to CRC Press, Taylor & Francis Group for their exceptionally kind support, which made our efforts successful.

Tanveer Bilal Pirzadah and Reiaz Ul Rehman

Acknowledgments

The present book has been a significant research challenge and was completed with the support of many individuals. Here I express my gratitude to a few of those people.

I would like to express my sincere gratitude to my supervisor, Dr. Reiaz Ul Rehman, for his encouragement and support.

I would like to express my deepest gratitude and sincerest appreciation to Prof. Inayatullah Tahir, Dr. Mohammed Irfan Qureshi and Dr. Sanjeet Singh for their encouragement and support.

Special thanks to my dearest wife, Dr. Bisma Malik, for her help, support and contributions.

I am highly thankful to Prof. Ivan Kreft, president of the International Buckwheat Research Association (IBRA), Ljubljana, Slovenia, for writing a foreword for the book.

I am very grateful to my parents for their endless encouragement, understanding and patience. They have been a constant source of support both emotionally and financially. Special appreciation to my family members Arjumand, Shahnaz, Waseem and Zahoor Ahmad and my little angels Emaan, Enaya, Shahzaib and Shah Hamdaan for their love and support.

We are very sure that this book will interest scientists, graduates, undergraduates and postdocs who are investigating "underutilized crops," particularly buckwheat crop improvement programs.

Dr. Tanveer Bilal Pirzadah
Assistant Professor
University Centre for Research and Development (UCRD)
Chandigarh University, Punjab, India

Authors

Dr. Tanveer Bilal Pirzadah is Assistant Professor at University Centre for Research and Development (UCRD), Chandigarh University, Punjab, India. After completing his doctorate at the University of Kashmir, Srinagar, India, he worked as an Assistant Professor at the Cluster University Srinagar, India. Dr. Tanveer has about eight years of research and teaching experience in bioresources management, biofuels, plant stress physiology and biotechnology. He is the recipient of the Young Scientist Award (RA-VD00921). He is also a fellow of the Plantae group of the American Society of Plant Biologists and a member of the World Academy of Sciences. Currently, Dr. Tanveer is engaged in studying the agronomical, physiological and biochemical and nutritional aspects in underutilized crops including buckwheat. He has to his credit 31 research articles published in peer-reviewed international journals, 30 book chapters in edited volumes and three books with international publishers.

Dr. Reiaz Ul Rehman is Assistant Professor in the Department of Bioresources, School of Biological Sciences, University of Kashmir, Srinagar, India. He is actively involved in teaching Master's and Integrated PhD students and has guided several M.Sc., M.Phil and PhD theses. He is the recipient of several fellowships at both national and international levels, including the Marie Curie IIF award (MC-IIF-FP7-219339). His research interests include the agronomical, physiological and biochemical aspects of and mineral nutrition in underutilized crops including buckwheat. He has to his credit two published books, 40 articles in various journals and 25 book chapters in several edited volumes.

1 Origin and History of Buckwheat

INTRODUCTION

The name *Fagopyrum* is derived from the Greek word *fagos* meaning "beech" and *pyrum* meaning "wheat," and the common name, buckwheat, is from *Buchswein*, German for beech-wheat because of the triangular seeds that look like little beech-nuts (Paulickova 2008). However, it is neither a nut nor a cereal but is categorized as a separate group called pseudocereals because it has both similarities and dissimilarities with cereals. Further, it is not considered a true cereal because cereals are monocots, unlike buckwheat, which is dicot and contains a cereal-like starchy endosperm. Reports have also revealed that, depending on their botanical origin, the starch granules of buckwheat differ in size, shape and composition (Farooq et al. 2016). The morphology of buckwheat shows that it is an annual, broad-leafed plant and attains a maximum height of about 0.6–1.5 meters (Skrabanja et al. 2004). Buckwheat belongs to the family *Polygonaceae* and genus *Fagopyrum*, which comprises approximately 34 species including both diploids (2n = 16) and tetraploids (2n = 4x= 32) (Zhou et al. 2018; Singh et al. 2020); likewise, some new species have recently been included, namely *Fagopyrum crispatifolium* (Liu et al. 2008), *Fagopyrum pugense* (Tang et al. 2010), *Fagopyrum wenchuanense* (Shao et al. 2011) and *Fagopyrum qiangcai* (Shao et al. 2011), which are well characterized both taxonomically as well as phylogenetically (Zhou et al. 2012). Further, *Fagopyrum* is categorized into two groups, i.e. the cymosum group, which has big lusterless achene partly enclosed with insistent perianths, and consists of 8 species; and the urophyllum group, which has small lustrous achene thoroughly enclosed with insistent perianths and constitutes approximately 16 species (Ohnishi and Matsuoka 1996; Ohnishi 2016; Zhou et al. 2018). *F. esculentum*, *F. tataricum*, *F. homotropicum* and *F. cymosum* are included in the cymosum group along with the allotetraploid perennial species, i.e. *F. cymosum* and its two diploid varieties, *F. cymosum* var. pilus and *F. cymosum* var. megaspartanium of the cymosum group, forming the *cymosum* complex (Chen 1999) (Figure 1.1). Ohnishi (2016) reported that these can readily cross within the complex and are closely related to each other. Data has also revealed that *F. megaspartanium* and *F. pilus* of the cymosum complex might be the ancestors of the two commonly cultivated buckwheat species, i.e. *F. tataricum* and *F. esculentum*, based on certain parameters such as the meiotic behavior of hybrid chromosome, DNA polymorphism and allozyme analyses (Chen 2001; Ohsako and Ohnishi 2000; Li et al. 2013; Chen et al. 2004). In Jammu and Kashmir, four different *Fagopyrum* species are cultivated, viz. *F. esculentum*

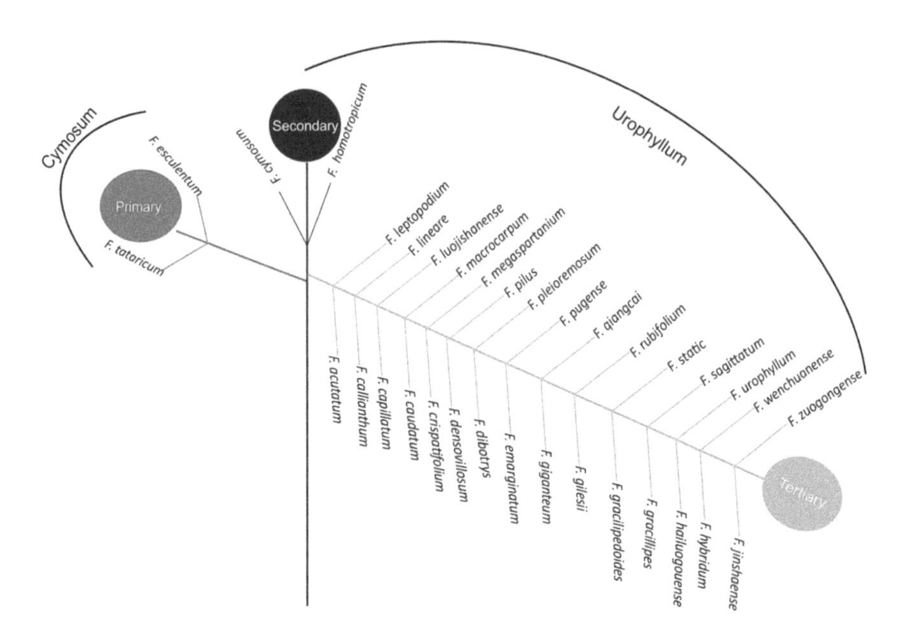

FIGURE 1.1 Classification of genus *Fagopyrum* on the basis of distinct morphological identification that forms the base of gene pool classification.

Moench, *F. tataricum* Gaertn., *F. sagittatum* Gilib. and *F. kashmirianum* Munshi (Munshi 1982; Tahir and Farooq 1988) and one species, viz. *F. cymosum* Meissn, which grows wild and is propagated through rhizome (Tahir and Farooq 1989). Moreover, the species of buckwheat that have been found so far are diploid (2n = 16), except *F. cymosum* and *F. gracilipes*, which are tetraploid in nature (4n = 32). Buckwheat is characterized by a single, upright and hollow stem that exhibits color variation ranging from green to red and turns brown at maturity. The shape of the leaves is heart-shaped or triangular, 5–7 cm long, and is arranged alternately on the stem. The flower is attractive and is usually clustered in racemes at the tip of branches or on short pedicles emerging from the leaf axils, and the inflorescence is composed of 7–9 blossoms. Moreover, the flowers are bisexual in nature with apparent parts that include 3–5 sepals and petals (tepals), 6–9 stamens and a single pistil, and the color of the flower ranges from white or light green to pink or red (Cawoy et al. 2009). The ovary is superior and possesses one locule with an egg that develops into an achene, and the base of the ovary is associated with nectar-secreting glands. One of the specialized properties of buckwheat is that it exhibits sexual dimorphism, i.e. it comprises of two types of flowers, viz. pin flowers consisting of a short stamen and long pistil and thrum flowers with a long stamen and short pistil. Depending upon the variety, the shape of the seeds varies from triangular to wing-shaped, and the texture of the seeds ranges from glossy brown or black to silvery gray (Krkoskova and Mrazova 2005) (Figure 1.2). The seed comprises an outer thick covering (hull) with a lesser density than water and an inner part (groat) that resembles a cereal kernel in its raw

FIGURE 1.2 Images of buckwheat, showing the growth habit of cultivated buckwheat (*Fagopyrum esculentum*) and the seed (A and B), with a comparison of the seed of Tartary buckwheat (*Fagopyrum tataricum*) (C and D).

chemical composition and structure. Li and Zhang (2001) reported that the stiffness of the hull depends upon the cultivar; for example, *F. esculentum* has a softer scale than its relative species *F. tataricum*. In addition, buckwheat has a dense and fibrous root system with a deep taproot that makes up about 3–4% of the total plant weight (Woo et al. 2016); this plays a pivotal role in various physiological processes such as promoting earthworm activity and thus enhancing the porosity of the soil (Farooq et al. 2016).

ORIGIN AND DISTRIBUTION

Buckwheat is believed to have originated in central Asia, but the origins of its domestication date back 4,000–5,000 years in southern China (Gondola and Papp 2010). Li and Yang (1992) reported that in China, the earliest findings of buckwheat came from a tomb of the Western Han Dynasty, where the seeds of common buckwheat were dated from the 1st to the 2nd century and thus is regarded as the original center of buckwheat. The history of buckwheat revealed that it originated in central Asia where it was widely cultivated as a main or supplementary food, and it was brought to Europe during the Middle Ages. Since the 1980s, farmers and buckwheat researchers have identified various wild buckwheat species in the south-western part of China and have developed some new perspectives based on these findings. Jiang and Jia (1992) proposed that the Daliangshan region was amongst the places of origin of Tartary buckwheat, based on the large number of wild buckwheat types, particular ecological circumstances, and traditional practices related to buckwheat. On the basis of research findings on buckwheat and the national history of Yunnan, it originated in south-western China, on the eastern side of the Himalayas, as reported by Li and Yang (1992). More precisely, Yunnan, Daliangshan, Xiaoliangshan and the border region of Yunnan and Sichuan were the center of the origin of buckwheat. Ye and Guo (1992) reported that from the botanical point of view, the south-western part of China was not only the differentiation and distribution center, but also the original location of buckwheat. It was reported that the wild relatives of *F. esculentum* ssp. Ancestral were the ancestors of cultivated species *F. esculentum* Moench, based on investigating morphology, reproductive biology and isozyme, and the marker (RAPD and AFLP) analysis (Ohnishi 1995, 1998a, 1998b, 2004). Likewise, another relationship was investigated between the related wild species *F. tataricum* ssp. potanini and the cultivated species *F. tataricum* Gaertn; isozyme and genetic analysis (RAPD and AFLP) revealed that Tartary buckwheat originated in the eastern part of Tibet and the common area of Yunnan and Sichuan (Tsuji and Ohnishi 2000, 2001a, 2001b). Thus, due to the widespread nature of wild buckwheat species, cultivation history, phylogenetic analysis, historical evidences and molecular studies confirmed that southwest China was not only the distribution center but the center of origin of its cultivated species. After the domestication of buckwheat in southwest China, it later spread into other areas through two routes: one from southwest to northern China, then to the Korean peninsula and later to Japan; and the other from China via Tibet to Bhutan, Nepal and India, then via Kashmir to Poland (Murai and Ohnishi 1995). One of the oldest written references in which buckwheat is mentioned was in Klaret's *Versed Glossary*, which dates from around the year 1365 AD. It was imported into European countries during 13th century via the "Silk Road" during invasions by Mongolian, Turkish and other troops. It was easy for the soldiers to transport buckwheat seeds and prepare nutritional foods. Later it was introduced into central Europe and Bohemian lands probably via two separate routes – either outside the Carpathians via Polish territory or the Danubian region (Pirzadah et al. 2020). Buckwheat was the most important food in Japan in 800 AD. It spread to Europe via Siberia and southern Russia from 1200 to 1300 AD. Countries like Ukraine, Germany and Slovenia were possibly the first countries to grow buckwheat in Europe, and then Belgians, French, Italians and English started growing buckwheat in the 17th century. After the 17th century, buckwheat was brought to America by the Dutch and nowadays buckwheat is widely cultivated on a large scale around the world in many countries. Among Asian countries, buckwheat is cultivated in Pakistan, India, China, Nepal, Bhutan, Mongolia, North Korea, far eastern Russia and Japan. Recently, Japan has increased the cultivated areas of common buckwheat, particularly in transformed paddy fields (MAFF 2019). Buckwheat cultivation also takes place in Europe, the United States, Canada, Australia and South Africa (Campbell 1997). Buckwheat is considered a young grain crop in Europe because it was not known at all during the Middle Ages (Figure 1.3).

In India it was first cultivated in the Himalayan region and is commonly known as *Kuttu*, *Phaphra* in Hindi and in Kumaon as *Ogal* (Sonia et al. 2012). Buckwheat is generally consumed by the Hindus on particular fasting days, especially during *Navaratri*. The crop is widely grown in the high mountains of Jammu and Kashmir in the west to Arunachal Pradesh in the east (Joshi 1999). It is densely cultivated in high altitudinal regions of Jammu and Kashmir, Himachal Pradesh, Uttrakhand, West Bengal, Sikkim, Meghalaya, Arunachal Pradesh and Manipur (Anonymous 1987–2001) (Figure 1.4). However, its cultivation has totally been eradicated especially in some hilly regions of northwest India; moreover, the situation in the trans-Himalayan region of Himachal Pradesh is also not promising as the production yield of Lahaul and Spiti has been declining due to the diversification of the land use model in favor of off-season crops such as vegetables, hops and some medicinal plants like *Saussaurea lappa* and *Inula racemosa*. In Kashmir, buckwheat is commonly known as *trumba shirin* (Ahmed et al. 2013). During the 19th century, the first data relating to buckwheat cultivation in Kashmir was described by Lawrence in 1895. Gohil, (1984) reported that buckwheat was introduced into Kashmir probably around 1200 AD. Lawrence describes buckwheat as an important and hardy crop. People in the higher reaches of Kashmir often cultivated buckwheat as a substitute for rice because it was suitable for earning a living. In different regions of Ladakh, buckwheat is known by various names such as *dyat*, *dro*, *bro* and *fafar*. In Kargil, generally two variants of buckwheat are grown, which include the yellow-colored, small-sized *brosuk* and the black-colored, larger-sized *gyamrus*. A few years ago, buckwheat was one of the staple foods of Ladakh. *Kiseer* or *giziri*, which is actually similar to the plain *dosa*, is one of the famous dishes prepared from buckwheat flour in

FIGURE 1.3 World production sites of buckwheat.

FIGURE 1.4 Buckwheat production sites in India.

the Kargil district of Ladakh division. Recently, the cultivation and consumption of buckwheat has been rapidly declining. In 1997, the organization *Wir laden zumHad´n*, meaning "we invite you to buckwheat," was established by various stakeholders such as farmers, traders and the village community Schwabegg. Currently many researchers all over the world focus on buckwheat cultivation due to its tremendous nutritional and medicinal properties.

BUCKWHEAT GERMPLASM RESOURCES AROUND THE WORLD

Due to the long history of buckwheat cultivation and its spread worldwide, it thus can be found on all continents except Antarctica. FAO (2020) reports that 25 countries are involved in buckwheat cultivation and the total area under buckwheat cultivation is about 3,001,495 ha, which produces 2,905,294 metric tonnes. The main cultivating areas of buckwheat are Russia, China, Ukraine, the United States, Kazakhstan, Poland, Japan, Brazil, Lithuania, France, Tanzania, Belgium, Nepal, Latvia, Bhutan, South Korea, Slovenia, Czech Republic, Estonia, Bosnia, South Africa, Hungary, Georgia and Kyrgyzstan. Currently, China is the leader in buckwheat production followed by Russia with 9,79,914 ha under cultivation, contributing about 892,322.58 metric tonnes (FAOSTAT 2020) (Figure 1.5). However, countries like Ukraine, the United States, Kazakhstan, Poland, Japan, Brazil, Lithuania and France contribute an area of approximately 300,000 ha for buckwheat cultivation. The top ten producers of buckwheat production in the world are represented in Figure 1.6. These countries possess historic evidence of buckwheat cultivation and thus could be considered as germplasm repositories. Moreover, 2,000 samples of buckwheat germplasm have been collected by various organizations such as the Chinese Academy of Agricultural Sciences and Russia's All-Russian Plant Research Institute. The aforementioned countries predominantly focus on the cultivation of *F. esculentum*, and only a few nations like Nepal, Bhutan, Pakistan and India grow *F. esculentum* and *F. tataricum* simultaneously. The largest area for the cultivation of *F. tataricum* is in China and occupies about 250,000 hm^2 each year. Moreover, due to the rich diversity of wild buckwheat species in China, it has generally been dispersed in the southern provinces, viz. Guizhou, Sichuan, Yunnan, Shaanxi, Chongqing, Tibet, Hunan, Hubei Zhejiang, Anhui, Fujian, Guangdong, Guangxi and Hainan, and specifically in south-western China, including Yunnan, Sichuan, Guizhou, Chongqing and Tibet; all wild species are restricted to this region, so this region has not only been regarded as the distribution and center of diversity for buckwheat, but also as the birthplace of cultivated species (common buckwheat and Tartary buckwheat) (Figure 1.7).

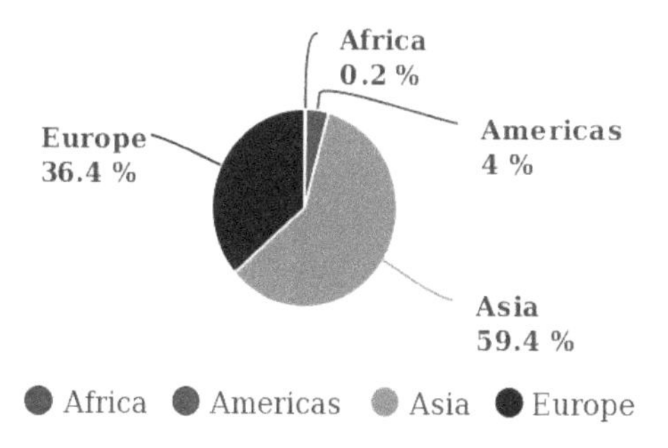

FIGURE 1.5 Average production share of buckwheat by region (1961–2028) (Source: FAOSTAT, November 6. 2020).

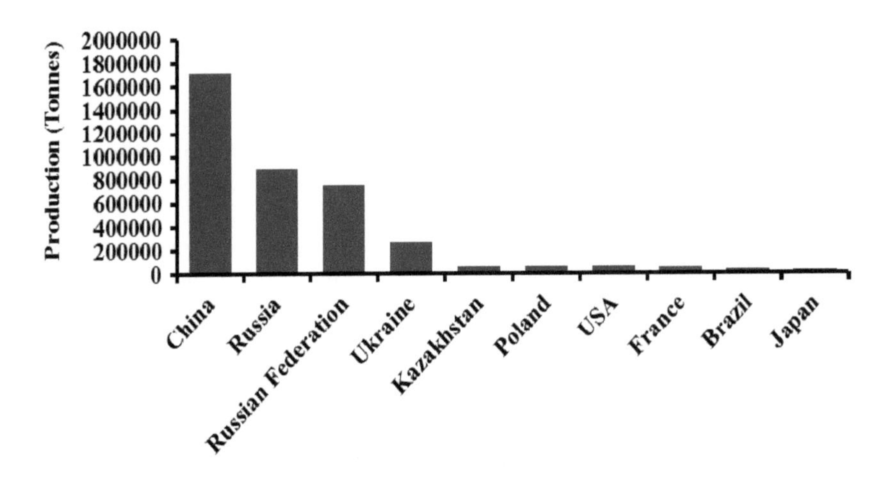

FIGURE 1.6 Graphical representation of the top 10 buckwheat production countries (FAOSTAT November 6, 2020).

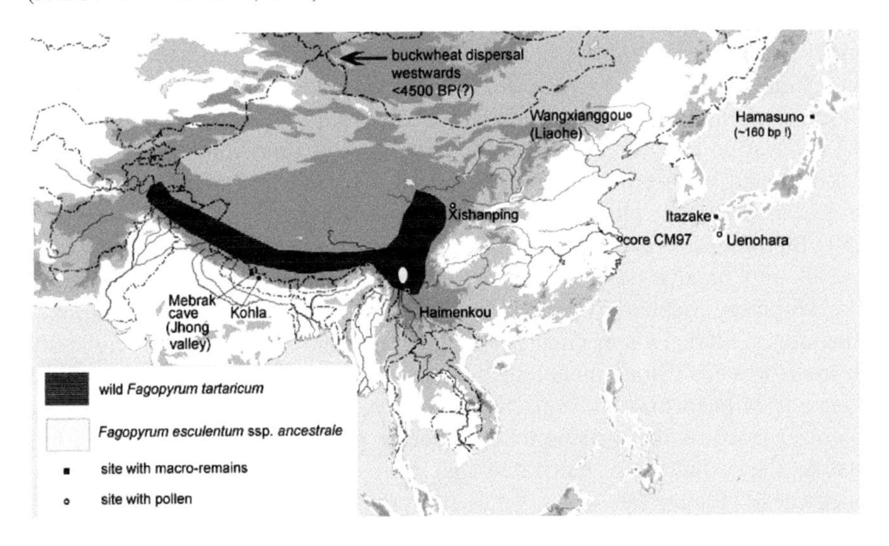

FIGURE 1.7 Map showing distribution of wild *F. tataricum*, the subzone of wild *F. esculentum*, and archaeological and palynological sites that have produced early evidence for buckwheat.

PRESENT STATUS OF BUCKWHEAT GERMPLASM RESEARCH

In the 1980s, the collection of buckwheat germplasm was begun by various recognized centers. With the help of the International Board for Plant Genetic Resources (IBPGR), scientists explored and collected more than 10,000 samples of wild buckwheat germplasm in the Himalayan region, of which half of were contributed by South and East Asia (Table 1.1) (Campbell 1997; Joshi 1999;

TABLE 1.1
Buckwheat Germplasm Conserved in Major Gene Banks Globally

Institute/organization	Country	Total germplasm holding	Major species represented	References
Institute of Crop Sciences, Chinese Academy of Agricultural Sciences (CAAS)	China	2,804	*F. esculentum* and *F. tataricum*	Zhou et al. (2018)
N I Vavilov All-Russian Institute of Plant Industry	Russia	2,116	Landraces, cultivars, and wild forms of *F. esculentum* and *F. tataricum* subsp. *multholium*	Zhou et al. (2018)
V Y Yuriev Institute of Plant Production	Ukraine	1,600	Landraces, cultivars and wild forms of *F. esculentum* and *F. tataricum*	Zhou et al. (2018)
Podillya State Agricultural University	Ukraine	900	*F. esculentum* and *F. tataricum*	Zhou et al. (2018)
National Bureau of Plant Genetic Resources	India	1,050	*F. esculentum*, *F. tataricum*, *F. sagittatum*, *F. cymosum*, *F. tataricum* ssp. *Himalianium* and *F. esculentum* ssp. *emerginatum*	Rana et al. (2016)
National Agriculture Genetic Resource Centre	Nepal	511	Indigenous landraces *F. esculentum* and *F. tataricum*	Paudel et al. (2016)
University of Ljubljana	Slovenia	378	Landraces of *F. esculentum,* accessions of *F. tataricum*	Zhou et al. (2018)
National Institute of Agro-Biological Sciences	Japan	226	Landraces of *F. esculentum* and varieties of *F. tataricum*	Katsube-Tanaka (2016)
Crop Research Institute	Prague	170	Mostly exotic accessions of *F. esculentum*	Cepkova et al. (2009)
Crop Research Institute	US	161	*F. esculentum* and *F. tataricum*	Zhou et al. 2018
Canada Research Station	Canada	572	*F. esculentum* and *F. tataricum*	Zhou et al. 2018

Chauhan et al. 2010; Zhou et al. 2018). These germplasm assets were preserved in long- (–20°C) or medium-term storage conditions (5°C, RH40%). This repository of buckwheat germplasm not only forms a foundation for genetics and genetic engineering but is also a valuable asset to taxonomists and phylogenetic researchers as well as other buckwheat-related scientists. Wild species of buckwheat contain enormously valuable genes that could play an essential role not only in enhancing the nutrition profile but also in various other crop improvement programs (Dar et al. 2018). Hence, buckwheat research needs to take a collaborative approach in order to unravel the novel traits that will play a pivotal role in various breeding programs.

CONCLUSION AND FUTURE POTENTIAL

In conclusion, buckwheat is an important underutilized crop with strong potential for domestication. Enhancement of the extent of its incorporation into the agricultural portfolio through appropriate species-specific interventions targeted at problems that hinder its domestication would be an important step towards its domestication and greater agricultural diversification.

REFERENCES

Ahmed, A., Khalid, N., Ahmad, A., Abbasi, N. A., Latif, M. S. Z. and Randhawa, M. A. 2013. Phytochemicals and biofunctional properties of buckwheat: A review. *J Agric Sci* 1–21. doi:10.1017/S0021859613000166.

Anonymous 1987–2001. *Annual Report, NBPGR Regional Station*. Barapani, Meghalaya.

Campbell, C. G. 1997. *Buckwheat: Fagopyrum esculentum Moench. Promoting the Conservation and Use of Underutilized and Neglected Crops, 19*. International Plant Genetic Resources Institute, Rome. ISBN:92-9043-345-0.

Cawoy, V., Ledent, J. F., Kinet, J. M. and Jacquemart, A. L. 2009. Floral biology of common (*Fagopyrum esculentum* Moench). *Eur J Plant Sci Biotechnol* 3: 1–9.

Cepkova, P. H., Janovska, D. and Stehno, Z. 2009. Assessment of genetic diversity of selected Tartary and common buckwheat accessions. *Span J Agric Res* 7(4): 844–854.

Chauhan, R., Gupta, N., Sharma, S. K., Sharma, T. R., Rana, J. C. and Jana, S. 2010. Genetic and genome resources in Buckwheat-present and future perspectives. *Eur J Plant Sci Biotechnol* 4: 33–44.

Chen, Q. F. 1999. A study of resources of *Fagopyrum* (Polygonaceae) native to China. *Bot J Linn Soc* 130(1): 53–64.

Chen, Q. F. 2001. Karyotype analysis of five buckwheat species (*Fagopyrum*) native to China. *Guihaia* 21: 107–110.

Chen, Q. F., Hsam, S. L. K. and Zeller, F. 2004. A study of cytology, isozyme and interspecific hybridization on the big-achene group of buckwheat species (*Fagopyrum*, Polygonaceae). *Crop Sci* 44(5): 1511–1518.

Dar, F. A., Pirzadah, T. B., Malik, B., Tahir, I. and Rehman, R. U. 2018. Molecular genetics of buckwheat and its role in crop improvement. In: Zhou, M., Kreft, I., Tang, Y. and Suvorova, G. (eds.) *Buckwheat Germplasm in the World*, 1st edn. Elsevier Publications, USA, pp. 271–286.

FAO 2020. http://www.fao.org/faostat/en/#data/QC/visualize.

Farooq, S., Rehman, R. U., Pirzadah, T. B., Malik, B., Dar, F. A. and Tahir, I. 2016. Cultivation, agronomic practices and growth performance of buckwheat. In: Zhou, M., Kreft, I., Woo, S. H., Chrungoo, N. and Wieslander, G. (eds.) *Molecular Breeding and Nutritional Aspects of Buckwheat.* Elsevier Inc., Amsterdam, Boston. pp. 299–320.

Gohil, R. N. 1984. Buckwheat in India – Past, present and future. *Fagopyrum* 4: 3–6.

Gondola, I. and Papp, P. P. 2010. Origin, geographical distribution and phylogenic relationships of common buckwheat (*Fagopyrum esculentum* Moench.). In: Dobranszki, J. (eds.) Buckwheat 2. *Eur J Plant Sci Biotechnol* 4(2): 17–32.

Jiang, J. F. and Jia, X. 1992. Daliangshan region in Sichuan Province is one of the habitats of tartary buckwheat. In: *Proceedings of 5th International Symposium on Buckwheat at Taiyuan, China.* Agricultural Publishing House, pp. 17–18.

Joshi, B. D. 1999. Status of buckwheat in India. *Fagopyrum* 16: 7–11.

Katsube-Tanaka, T. 2016. Buckwheat: Production, consumption and genetic resources in Japan. In: Zhou, M., Kreft, I., Woo, S. H., Chrungoo, N. and Wieslander, G. (eds.) *Molecular Breeding and Nutritional Aspects of Buckwheat.* Academic Press, Cambridge, pp. 61–80.

Krkoskova, B. and Mrazova, Z. 2005. Prophylactic components of buckwheat. *Food Res Int* 38(5): 561–568.

Li, F. L., Zeller, F. J., Huang, K. F., Shi, T. X. and Chen, Q. F. 2013. Improvement of fluorescent chromosome *in situ* PCR and its application in the phylogeny of the genus *Fagopyrum* Mill. using nuclear genes of chloroplast origin (cpDNA). *Plant Syst Evol* 299(9): 1679–1691.

Li, Q. Y. and Yang, M. X. 1992. Preliminary investigation on buckwheat origin in Yunnan, China. In: *Proceedings of 5th International Symposium Buckwheat at Taiyuan.* Agricultural Publishing House, pp. 44–46.

Li, S. and Zhang, Q. H. 2001. Advances in the development of functional foods from buckwheat. *Crit Rev Food Sci Nutr* 41(6): 451–464.

Liu, J. L., Tang, Y., Xia, Z. M., Shao, J. R., Cai, G. Z., Luo, Q. et al. 2008. *Fagopyrum crispatifolium* Liu JL, a new species of Polygonaceae from Sichuan China. *J Syst Evol* 46: 929–932.

MAFF 2019. Planted area and production of buckwheat in 2018. Retrieved from http://www.maff.go.jp/j/tokei/kouhyou/sakumotu/sakkyou_kome/index.html.

Munshi, A. H. 1982. A new species of *Fagopyrum* from Kashmir Himalaya. *J Econ Tax Bot* 3: 627–630.

Murai, M. and Ohnishi, O. 1995. Diffusion routes of Buckwheat cultivation in Asia revealed by RAPD markers. Current advances in Buckwheat research. In: *Proceedings of 6th International Symposium on Buckwheat in Shinshu, Japan*, pp. 163–173.

Ohnishi, O. 1995. Discovery of new *Fagopyrum* species and its implication for the studies of evolution of Fagopyrum and of the origin of cultivated buckwheat. In: Matano, T. and Ujihara, A. (eds.. *Current Advances in Buckwheat Research. Vol. I–III: Proceedings of 6th International Symposium on Buckwheat in Shinshu*, August 24–29. Shinshu University Press, pp. 175–190.

Ohnishi, O. 1998a. Search for the wild ancestor of buckwheat. I. Description of new Fagopyrum species and their distribution in China. *Fagopyrum* 18: 18–28.

Ohnishi, O. 1998b. Search for the wild ancestor of buckwheat. III. The wild ancestor of cultivated common buckwheat, and of Tartary buckwheat. *Econ Bot* 52(2): 123–133.

Ohnishi, O. 2004. On the origin of cultivated buckwheat. In: *Proceedings of 9th International Symposium on Buckwheat in Prague*, pp. 16–21.

Ohnishi, O. 2016. Molecular taxonomy of the genus *Fagopyrum.* In: Zhou, M., Kreft, I., Woo, S. H., Chrungoo, N. and Wieslander, G. (eds.) *Molecular Breeding and Nutritional Aspects of Buckwheat.* Academic Press, Cambridge, pp. 1–12.

Ohnishi, O. and Matsuoka, Y. 1996. Search for the wild ancestor of buckwheat. II. Taxonomy of *Fagopyrum* (Polygonaceae) species based on morphology, isozymes and cpDNA variability. *Genes Genet Syst* 72(6): 383–390.

Ohsako, T. and Ohnishi, O. 2000. Intra- and interspecific phylogeny of wild Fagopyrum (Polygonaceae) species based on nucleotide sequences of noncoding regions in chloroplast DNA. *Am J Bot* 87(4): 573–582.

Paudel, M. N., Joshi, B. K. and Ghimire, K. H. 2016. Management status of agriculture plant genetic resources in Nepal. *Agron JN* 4: 74–90.

Paulickova, I. 2008. Rutin an Effective Component of Functional Foods. Food Research Institute, Prague, pp. 1–7.

Pirzadah, T. B., Malik, B., Tahir, I. and Rehman, R. U. 2020. Buckwheat journey to functional food sector. *Curr Nutr Food Sci* 16(2): 134.

Rana, J. C., Singh, M., Chauhan, R. S., Chahota, R. K., Sharma, T. R., Yadav, R. and Archak, S. 2016. Genetic resources of buckwheat in India. In: Zhou, M., Kreft, I., Woo, S. H., Chrungoo, N. and Wieslander, G. (eds.) *Molecular Breeding and Nutritional Aspects of Buckwheat.* Academic Press, Cambridge, pp. 109–135.

Shao, J. R., Zhou, M. L., Zhu, X. M., Wang, D. Z. and Bai, D. Q. 2011. *Fagopyrum wenchuanense* and *Fagopyrum qiangcai*, two new species of Polygonaceae from Sichuan, China. *Novon* 21(2): 256–261.

Singh, M., Malhotra, N. and Sharma, K. 2020. Buckwheat (*Fagopyrum* sp.) genetic resources: What can they contribute towards nutritional security of changing world? *Genet Resour Crop Evol* doi:10.1007/s10722-020-00961-0.

Skrabanja, V., Kreft, I., Golob, T., Modic, M., Ikeda, S., Ikeda, K., Kreft, S., Bonafaccia, G., Knapp, M. and Kosmel, K. J. 2004. Nutrient content in buckwheat milling fractions. *Cereal Chem* 81(2): 172–176.

Sonia, M., Gupta, D. and Gupta, R. K. 2012. Evaluation of nutritional and antioxidant potential of Indian Buckwheat grains. *Indian J Trad Knowl* 11: 40–44.

Tahir, I. and Farooq, S. 1988. Review article on buckwheat. Buckwheat Newsletter. *Fagopyrum* 8: 33–53.

Tahir, I. and Farooq, S. 1989. Grain and leaf characteristics of perennial buckwheat (*F. cymosum* Meissn). Buckwheat Newsletter. *Fagopyrum* 9: 41–43.

Tang, Y., Zhou, M. L., Bai, D. Q., Shao, J. R., Zhu, X. M., Wang, D. Z. and Tang, Y. 2010. *Fagopyrum pugense* (*Polygonaceae*), a new species from Sichuan, China. *Novon* 20(2): 239–242.

Tsuji, K. and Ohnishi, O. 2000. Origin of cultivated Tartary buckwheat (*Fagopyrum tataricum* Gaertn.) revealed by RAPD analyses. *Genet Resour Crop Evol* 47(4): 431–438.

Tsuji, K. and Ohnishi, O. 2001a. Phylogenetic position of east Tibetan natural populations in Tartary buckwheat (*Fagopyrum tataricum* Gaertn.) revealed by RAPD analyses. *Genet Resour Crop Evol* 48(1): 63–67.

Tsuji, K. and Ohnishi, O. 2001b. Phylogenetic relationships among wild and cultivated Tartary buckwheat (*Fagopyrum tataricum* Gaertn.) populations revealed by AFLP analyses. *Genes Genet Syst* 76(1): 47–52.

Woo, S. H., Roy, S. K., Kwon, S. J., Cho, S. W., Sarker, K., Lee, M. S., Chung, K. Y. and Kim, H. H. 2016. Concepts, prospects, and potentiality in buckwheat (*Fagopyrum esculentum* Moench): A research perspective. In: Zhou, M., Kreft, I., Woo, S. H., Chrungoo, N. and Wieslander, G. (eds.) *Molecular Breeding and Nutritional Aspects of Buckwheat*, Elsevier Inc., Amsterdam, Boston. pp. 21–49.

Ye, N. G. and Guo, G. Q. 1992. Classification, origin and evolution of genus Fagopyrum in China. In: *Proceedings of 5th International Symposium on Buckwheat at Taiyuan, China.* Agricultural Publishing House, pp. 19–28.

Zhou, M. L., Bai, D. Q., Tang, Y., Zhu, X. M. and Shao, J. R. 2012. Genetic diversity of four new species related to southwestern Sichuan buckwheat as revealed by karyotype, ISSR and allozyme characterization. *Plant Syst Evol* 298(4): 751–759.
Zhou, M. L., Kreft, I., Suvorova, G., Tang, Y. and Sun-Hee, W. (eds.) 2018. *Buckwheat Germplasm in the World.* Academic Press, Cambridge, p. 382.

2 Cultivation and Agronomic Practices of Buckwheat

INTRODUCTION

The adoption of mono-culturing techniques and partial technological development and its use are only practiced in the case of a less high energy demanding plant species. Therefore, global food security and economic growth depend on only a few crops that have jeopardized future food supply and rural livelihoods. Wheat, rice and corn are the major cereals that support more than 50% of the global calorie demand (Figure 2.1). While these grains are an essential part of many diets around the world, they lack substantial amounts of micronutrients and phytonutrients. As a result, it is estimated that micronutrient deficiencies affect an estimated 2 billion people worldwide, raising health concerns due to over-dependence on these major cereal crops (FAOSTAT 2013). The limited agricultural portfolio raises many genuine questions as to how efficiently newer crops could subsidize world food security, alleviating poverty and the hidden hunger crisis. The dependence on limited crops diminishes the capacity of farmers along with their immediate social ecosystem to acclimate to new environments for their needs and livelihood opportunities (FAOSTAT 2013). To address these concerns, it is necessary to broaden the slate of research and development with a much wider range of cultivated plant species. One of the means or ways to address the concerns of farmers and protect their livelihood is by enabling the growth of locally important plant species. By adopting this approach, many problems related to agricultural methodology and environmental management will be addressed. In addition, the loss of various crop varieties has prompted agricultural scientists around the globe to retrieve research data and disseminate knowledge on the production and utilization of various lost, underexploited and neglected crops. Supplementation of newer crops with pseudocereals provides a wider range of food alternatives on the shelves with better nutraceutical potential that would not only help to combat the hidden hunger crisis but also enhance the food basket in the future.

These crops are in high demand in the food, health and energy sectors, in addition to offering opportunities for job creation, especially for rural people. However, with the evolution of modern agricultural technologies, most traditional plant species have been neglected as they are held in low esteem, and some have

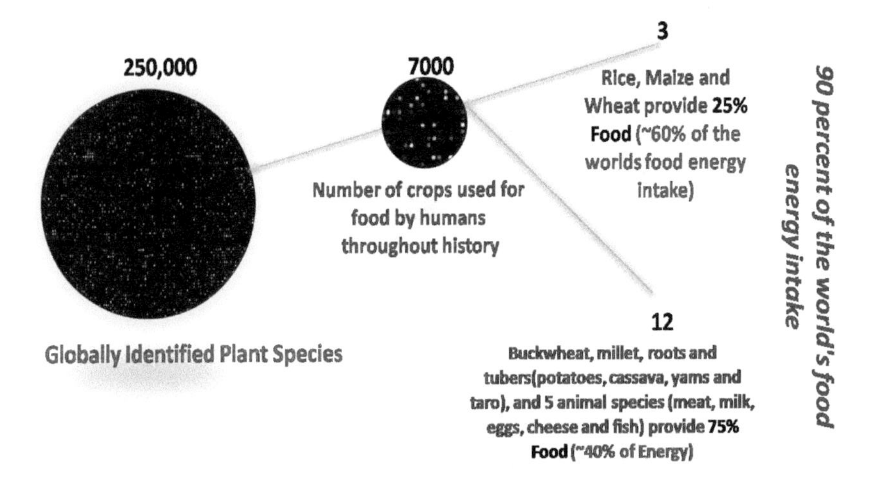

FIGURE 2.1 Diagrammatic representation of the contribution of major cereals and pseudocereals to the global food sector.

been neglected to the extent that there is severe erosion of their genetic makeup and hence they are often called "lost" crops.

Buckwheat is known as a versatile ancient crop in Asia and Central and Eastern Europe, which has been primarily grown for its grains (feed and food) and greens (fodder and vegetable) particularly in arid regions of the world (Figure 2.2). It thrives well in poor soil fertility conditions and is a multipurpose crop mainly grown for human consumption and livestock and poultry feed, as well as a green crop (Tsuzuki 2001), a soil-binding crop (Tundup et al. 2017) and a cover crop (Xuan and Tsuzuki 2004). It is from this perspective that buckwheat needs global attention, in particular from agronomists, molecular breeders and biotechnologists, to improve upon its limitations through agro-techniques for production/yield and for its revival to accelerate its mass cultivation.

AGRONOMIC PRACTICES

Preparation of Land and Sowing

Buckwheat is a short duration crop and thrives well in properly drained, sandy loam soils; it usually prefers moist and cool climatic conditions, and it is also well-adapted to acidic soils with pH < 5 (Olson 2001; Hore and Rathic 2002; Quinet et al. 2004; Jung et al. 2015; Pirzadah et al. 2019). Buckwheat does not require any sophisticated land preparation and has the capability to adapt well even in soils that are poorly tilled (Loch and Lazanyi 2010; Popovic et al. 2013a, 2013b; Ikanovic et al. 2013). However, it shows poor growth rate in saline and semi-arid regions (Farooq et al. 2016). Before sowing, land should be prepared in advance in order to enhance the porosity of the soil. There should be a proper drainage system to avoid seed submergence, as waterlogging will have a negative impact on

FIGURE 2.2 Photograph represents the cultivation practices of buckwheat in the valley of Kashmir. These pictures were taken by Dr. Tanveer Bilal and Dr. Reiaz Ul Rehman at different places in Gurez Valley, Jammu and Kashmir, India, in August 2013.

the seed germination process and ultimately will decrease the production yield. Seed sowing is one of the main criteria essential for production yield due to its important effects on the genotype of plant (Das et al. 2016). Thus, the correct sowing date gives maximal yield; moreover, the purpose of determining the sowing date is to find the proper time to plant cultivars so that the existing set of climatic conditions can be suitable for the germination process and for the survival of the plant (Hore and Rathic 2002; Jung et al. 2015). In the case of buckwheat, the growing season varies depending upon the agro-climatic conditions (Sobhani et al. 2014). The sowing season starts with the onset of the monsoon and continues up to August. In lower altitudinal areas, buckwheat is mostly sown from May to

August, while at higher altitudes it is usually sown from April to May (Ratan and Kothiyal 2011; Pirzadah et al. 2013). Moreover, it has also been reported that early June seeding produces the highest buckwheat yields. Lee et al. (2001) revealed that in China the best sowing time is more likely to be in late May, with a maximal yield of 2059 kg/ha, as compared to late April and early and late June. In European countries, the optimum sowing time is from mid-May to July, to avoid frosts (Bernath 2000; Halbrecq et al. 2005). The optimum sowing time for buckwheat in various regions of India is depicted in Table 2.1. Seeds should be sown at a depth of about 4–6 cm (Bjorkman 2012). However, seeds should be sown deeper during dry climatic conditions (Figure 2.3). Seeds should not be sown very deep because deep seeding delays emergence and decreases the uniformity of the crop,

TABLE 2.1
Sowing Time of Buckwheat in Different Parts of India

S. No.	Region	Sowing Time
1.	North-Western hills	June–July (rainy season) and March–April (spring season)
2.	North-Eastern hills including Assam	August–September, (October–November in Sikkim under intensive cropping systems)
3.	Nilgiri hills (Tamil Nadu)	April–May
4.	Plani hills (Tamil Nadu) and Kerala	January
5.	Chhattisgarh	September–November
6.	Jammu and Kashmir	March–April (Spring season) and September–October (Autumn season)

FIGURE 2.3 Peasant cultivation of Buckwheat, as shown in art. (a) *Sowing Buckwheat.* Source (public domain), (b) *Buckwheat Harvest* by Jean-François Millet, painted between 1868 and 1874, housed in the Museum of Fine Arts, Boston.

and the suggested sowing depth is approximately 4–6 cm; however under dry conditions, seeds should be sown very deep to get adequate moisture (Bjorkman 2012). The seeds are generally sown in rows (10 cm) apart or distributed randomly across the fields at a rate of 35–40 kg/ha when grown as a cereal crop, but at a rate of around 50 kg/ha when grown as a smothering/cover or fodder crop (Hore and Rathic 2002; Hulihalli and Shantveerayya 2018). In addition, the seeding rate should be restricted as it alters the growth of the plant, thus resulting in lower production yield, and the thinning may be started 20 days after sowing. Due to its allelopathic nature, the crop does not require weeding (Pirzadah et al. 2013; Babu et al. 2018). When buckwheat is being used as a cover crop, it should be sown at a higher rate in those fields where the growth rate of weeds is higher. However, in fields without weeds, seeds should be sown at a normal rate. It is recommended that 16 plants per square foot are suitable for optimum production. Due to the high degree of plasticity in buckwheat, its branches extensively recompense for lower plant populations. Buckwheat is a spring, thermophilous plant species. It is reported that in moist soil, buckwheat seeds begin to germinate at a temperature of 7–8°C. Studies also reported that the minimum temperature requirement for germination is 3–5°C. The seeds do not germinate when the temperature ranges from 0 to 4°C under field conditions (Ruszkowski 1980). It has been revealed that the optimum temperature for germination is 25–26°C, while other reports claim that buckwheat growth is faster at 20–25°C or 25–31°C (Ruszkowski 1980). It is very difficult to determine the maximum germination temperature for buckwheat; however, the upper temperature threshold lies between 35–45°C, while other reports contend that it is about 31°C (Ruszkowski 1980) because the seedlings wilt and the seeds stop absorbing water above this temperature limit.

Nutrient Requirements

The crop yield efficiency primarily depends on the soil nutrients, as analyzed by soil tests. Generally, buckwheat grows in less fertile soils and is thus considered a crop suitable for poor fertility soils (Radics and Mikohazi 2010; Rana et al. 2012; Pirzadah et al. 2013, 2020). Soils with high fertility are not suitable for the cultivation of buckwheat because it leads to lodging, and the crop is also unable to resist strong winds and heavy rains. Moreover, soils rich in nitrogen content are also unsuitable for the production of buckwheat as they decrease the amount of lignin content and other related enzymes, thus making the plant prone to lodging (Wang et al. 2015) as well as causing limited seed setting and delayed maturity. It has also been reported that buckwheat is often regarded as a phosphorus scavenger because it has the ability to utilize phosphorus efficiently from the soil (Figure 2.4). The phosphorus-scavenging efficiency of buckwheat results in early maturity and helps the plant to prevent lodging (Inamullah et al. 2012). Therefore, buckwheat requires both nitrogen and phosphorus in a specific ratio for efficient management (Farooq et al. 2016). Previous data also revealed that buckwheat has the capability to fix atmospheric nitrogen, which plays an important role in sustainable crop production (Alekseyeva 2002). Buckwheat has the unique ability to

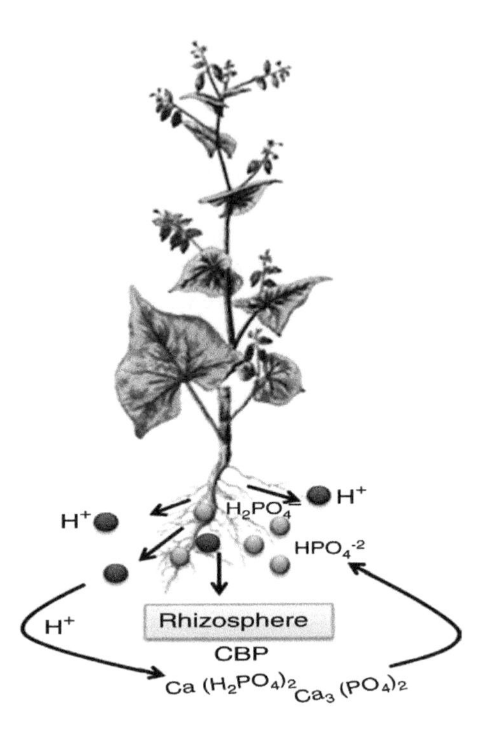

FIGURE 2.4 Buckwheat as an efficient phosphorus scavenger. Red dots indicate protons (H^+) secreted by buckwheat roots and green dots indicate forms of phosphorus ($H_2PO_4^-$) available to plants in a calcium-bound state, that is, calcium-bound phosphorus (CBP).

solubilize phosphorus, which otherwise may be unavailable to plants, by secreting a substance from the roots at the growing stage. Besides, the roots have also been found to have a high storage capacity for inorganic phosphate. Therefore, buckwheat can be used as a manure crop as it undergoes decomposition quickly, making phosphorus and other nutrients easily available to the succeeding crop. Buckwheat exhibits an excellent response to an even-handed fertilizer program, but it is not considered an intense nutrient user. The fertilizer recommended for growing buckwheat includes nitrogen (47 kg), phosphorus (22 kg) and potassium (40 kg) to harvest a yield of approximately 1600 kg/ha (Campbell and Gubbels 1978). In India, maximum yield is obtained upon applying nitrogen, phosphorus and potassium at the rate of 50:20:40 Kg or 1500–2000 kg farmyard manure/hectare (Phogat and Sharma 2000). In sandy loam soils it is suggested to add potash to the soil; moreover, in fields with high nitrogen levels, buckwheat shows lodging and ultimately there will be a decline in its production yield. The application of fertilizers greatly influences the grain yield; urea enhances the content of dry matter, proteins and soluble sugars, while ammonium nitrate increases the grain yield (Ganyushina and Lazarchik 1972). Rabe and Lovatt (1986) reported that increasing levels of ammonium nitrate reduces the arginine content of the grains. It has also been reported that buckwheat is often regarded as phosphorus scavenger because it has the ability to utilize phosphorus efficiently from the soil. Buckwheat has the unique feature of being able to solubilize phosphorus which otherwise may be unavailable to plants by secreting some substance from the roots at the growing stage. Moreover, the roots have also been found to have a high storage capacity for inorganic phosphate. Therefore, buckwheat can be used as a manure crop as it undergoes decomposition quickly, making phosphorus and other nutrients easily available to the succeeding crop. Further, it also suppresses weeds during growing season by releasing allelopathic compounds and thus is used for rejuvenating soils that are lacking in nutrition and have an annual weed problem (Kalinova et al. 2007). Buckwheat responds well to potassium fertilizers, especially in the form of sulfate rather than chlorides (Goncharik and Kozlova 1972). Moreover, buckwheat grows in soils with high aluminum content because of its ability to secrete the organic acid (oxalic acid) that causes chelation of the aluminum in the root rhizosphere (Pirzadah et al. 2019). Previous reports have also revealed that aluminum resistance in buckwheat could be also due to the immobilization and detoxification of aluminum by phosphorus in the root rhizosphere (Zhu et al. 2002; Ma et al. 1997). Buckwheat is considered an efficient green manure crop due to its fast growth rate; enrichment of soil nutrients, particularly nitrogen; phosphorus mobilization; fast decay rate; and better litter quality (Teboh and Franzen 2011). Thus, the incorporation of buckwheat into the soil modifies the health of the soil by improving the texture of the topsoil, improving its inclination and its porosity.

Pollination Management in Buckwheat

Buckwheat prefers cross-pollination, thus entomophily is an effective mode of pollination to produce the seed set. However, the degree of pollination efficiency depends upon various parameters, such as the morphology of flowers, insect abundance, nectar production, and transfer and deposition of pollen on an appropriate stigma (Woo et al. 2016; Farooq et al. 2016). Nectar production is influenced by the age of the plant, position of inflorescence, heteromorphy, ploidy level and environmental factors (Cawoy et al. 2008; Farooq et al. 2016; Dar et al. 2018). Moreover, buckwheat exhibits sexual dimorphism, i.e., common buckwheat produces two types of flowers: pins, which have a short stamen and long pistil; and thrums, which have a long stamen and short pistil. This dimorphic characterization makes common buckwheat self-incompatible (Figure 2.5). However, Tartary buckwheat has a homomorphic type of flower and thus possesses the ability to self-fertilize (Tahir et al. 1985; Woo et al. 2010, 2016; Farooq et al. 2016; Ruan et al. 2020). The various pollinators include Hymenoptera: bumble bees, honey bees, solitary bees and wasps; Diptera: Syrphidae, Calliphoridae and others; and Lepidoptera, Hemiptera, Nevroptera and other orders (Jacquemart et al. 2007). Moreover, the main pollinators of buckwheat belong to Apoidae (Hymenoptera, *Apis mellifera* and *Bombus* species) and Syrphidae (Diptera, *Eristalis* species) (Racys and Montviliene 2005; Pirzadah et al. 2013; Farooq et al. 2016). However, the honey bee is the most efficient pollinator as it visits both types of flowers (pins

FIGURE 2.5 Heteromorphic flowers of common buckwheat: (A) pin flower; (B) thrum flower; (C) long-homostyle flower; (D) short homostyle flower.

and thrums) for nectar collection; moreover, due to its foraging and prospecting behavior, it accelerates intermittent contacts with stigma (Bjorkman 1995; Jacquemart et al. 2007). As thrum-type flowers produce more nectar than pin types, the honey bee spends the majority of its time on thrum-type flowers. While the production of sugars is equal in both morphs, sucrose concentration is rich in thrum flowers (approximately 16.8% against 12.9%) and thus the sucrose/hexose ratio is dominant for thrum-type flowers (Cawoy et al. 2006). For the maximal production of buckwheat, one honey bee colony/acre is sufficient for effective pollination especially during sunny and warm days. It has been reported that the optimum temperature for effective pollination is 20°C (Sugawara 1956; Alekseyeva and Bureyko 2000; Farooq et al. 2016; Raun et al. 2020). Goodman et al. (2001) revealed that bumble bees and honey bees visit buckwheat mainly between 9 a.m. and 12 p.m. daily, while syphids are always active in the afternoon. The honey bee is the predominant pollinator, visiting approximately 14–20 flowers/minute, and it has been estimated that it works for about 4–5 h/day (Jacquemart et al. 2007). Jacquemart et al. (2007) conducted a cage experiment and concluded that the pollination efficiency of the honey bee is excellent as it deposited compatible pollen on the majority of the flowers (>90%) without discrimination among the floral forms. Furthermore, the rate of pollination is also affected by a low temperature as it reduces sucrose production and thus modifies the relative levels of sugars in buckwheat nectar. Reports have revealed that the rate of sugar secretion/flower can be approximately 15 times higher during cool and humid conditions than in dry climatic conditions. In addition, the optimum soil moisture for nectar secretion is on average 60%, but nectar secretion also declines under water stress conditions (Racys and Montviliene 2005).

Pest and Disease Management of Buckwheat

Generally buckwheat is resistant to severe pest infestation and diseases; however, there are some reports of pathogenic problems. Various scoop caterpillars such as *Euxoa tritici E. segetum*, *Trachea atriplicis*, *Phytometra gamma* and *Barathra brassicae* cause damage to buckwheat shoots, and some polyphytophages are responsible for causing damage to the root portion. Phytophages species like *Chaetocnema concinna* cause damage to stalks and young leaves. For some pests such as *Meligethes aeneus*, flowers also act as a forage material (Naumkin 2013). Other pests affecting buckwheat include wireworms, birds, rodents and aphids (Tahir et al. 1985; Bhat et al. 1986; Olson 2001). It has also been reported that cultivation of buckwheat earlier than the recommended time period makes it prone to aphid attacks, leading to a decline in crop production. These aphids secrete a greasy and sugary substance known as honeydew that encourages the growth of sooty mold on plants and thus causes damage to them. To avoid such infestations, a chemical (petroleum servo Agrospray at 7ml/L or neem oil 1500ppm at 3ml/L) can be used for the effective management of aphids (Babu et al. 2014, 2018). Jacquemart et al. (2012) also reported about 20 viruses (Tobacco Mosaic Virus-TMV; Cucumber Mosaic Virus-CMV) and a few bacteria associated with buckwheat. The severe pests of buckwheat include beetles and their larvae, which affect the shoots during early June. The larvae cause curvature of the stem and make the plant liable to lodging as well as causing severe damage to the nectiferous glands (Kuznetsova et al. 2012). A previous study revealed that a pest like *Rhinoncus sibiricus* causes approximately 30–50% reduction in the yield of *F. esculentum* (Kuznetsova and Klyka 2012; Klykov et al. 2014).

Crop Harvesting and Storage

As buckwheat exhibits an indeterminate growth habit, it is very difficult to determine the specific harvesting time period, so swathing is an essential step to minimize losses through shattering. However, swathing should be done in the early morning to minimize shattering losses (Babu et al. 2018). Campbell and Gubbels (1978) reported that shattering causes about 20–25% yield loss in buckwheat. Moreover, agro-climatic conditions also influence the harvesting period of buckwheat because at higher altitudes, harvesting is done late in the season, while at moderate and lower altitudes, harvesting is done very early. Under normal conditions, harvesting is done approximately ten weeks after planting. It has been observed that harvesting of buckwheat should be done when approximately 70–75% is matured and most of the leaves have been cast off. At this stage, threshing becomes an easy process as most of the seeds fall out quickly. Buckwheat should not be stored for long because of its susceptibility to rancidity. Before storage, seeds should be well dried, and the optimum moisture content and temperature for drying is approximately about 16% and 45°C respectively (Olson 2001; Small 2017; Farooq et al. 2016; Babu et al. 2018; Raun et al. 2020). For long-term storage, a moisture content of 13% is required while for short-term storage, a minimum temperature of 10°C and moisture content of <15% in grains is recommended (Olson 2001; Small 2017).

BUCKWHEAT AS A SMART COVER CROP IN AN INTERCROPPING SYSTEM

Buckwheat is regarded as a smart crop and is used in intercropping systems with soybeans, potatoes, solar hemp, millet, safflowers, sunflowers, etc. (Pirzadah et al. 2013; Pavek 2014). Most farmers use buckwheat as cover crop as it has the capability to soften the soil as well as enriching it with various nutrients (Petrich 2000). Moreover, it has the ability to control weeds; the pathway mechanism involves either shading or by means of chemical warfare known as an allelopathic effect, which makes it a promising biocontrol agent (Tahir and Farooq 1990; Geneau et al. 2011; Wirth and Gfeller 2016). It has been reported that buckwheat secretes a bioactive compound, diethyl phthalate, that causes the suppression of weeds around its periphery. Moreover, buckwheat possesses the ability to alter the soil's nitrogen content and makes it unavailable to weeds, thus inhibiting their growth. Kumar et al. (2008) also reported that once buckwheat is incorporated into the soil, it reduces the nitrogen level but at the same time enhances the potassium level, thus suppressing weed growth. Besides, buckwheat has a carbon: nitrogen ratio of 34, which immobilizes the nitrogen content during decomposition (Creamer and Baldwin 1999). Jacquemart et al. (2012) conducted a field study where 2 tons of buckwheat pellets were used in a rice field before planting, and they concluded that the production of rice was increased by 20% and the weed suppression by about 80%. In another study, buckwheat was found to control weeds approximately 75–99% by decreasing weed biomass when compared to bare soil treatments (Iqbal et al. 2003). Furthermore, compared to other crops such as clover, wheat, maize and sorghum, buckwheat is a rich source of calcium, magnesium, phosphorus, potassium and other micronutrients (zinc, iron, manganese, etc.) thus enriching the soil quality, as reported by various authors (Warman 1991; Pirzadah et al. 2018, 2020). Buckwheat is considered an efficient phosphorus scavenger, so the cultivation of buckwheat enriches the phosphorus content in the soil by about 10 times compared to wheat (Zhu et al. 2002). The mechanism behind the phosphorus scavenging activity involves buckwheat's ability to acidify the rhizosphere by draining protons (H^+) that in turn dissolve the calcium-bound phosphorus in an alkaline soil, making phosphorus available to the plant (Zhu et al. 2002).

CONCLUSION AND FUTURE POTENTIAL

Buckwheat is an underutilized crop that can be cultivated on marginal land and is projected as the golden crop of the future. It acts as a promising candidate in crop improvement programs as reported by various research and development (R&D) institutions, such as the International Plant Genetic Resources Institute (IPGRI) and the Consultative Group on International Agriculture (CGIAR). Current research must address important scientific issues related to buckwheat like the identification of desired agronomic and genetic traits pertaining to seed shattering, enhanced nutritional quality, self-compatibility and homostyly. In addition, interventions at the molecular level have immense potential to provide vital clues to tackle the inherent problems of buckwheat. Genetic improvement is crucial for determining the potential for rapid spread through seeds, propagules, etc.; in addition, it can help to understand various aspects of crop biology including the improvement of agricultural characteristics such as physiological and agronomic traits, abiotic or biotic tolerance and enhancement of nutraceutical and biofortification properties. As non-conventional biotechnological approaches are quite effective in enhancing desired traits, so they could therefore be used to solve the productivity and yield problems associated with buckwheat.

REFERENCES

Alekseyeva, E. S. 2002. Progress and prospects of buckwheat improvement in Ukraine – Current status and future research. *Fagopyrum* 19: 111–113.

Alekseyeva, E. S. and Bureyko, A. L. 2000. Bee visitation, nectar productivity and pollen efficiency of common buckwheat. *Fagopyrum* 17: 77–80.

Babu, S., Kalita, H., Singh, R., Gopi, R., Kapoor, C. and Das, S. K. 2014. Buckwheat (*Fagopyrum* spp.). In: Avasthe, R. K., Pradhan, Y. and Bhutia, K. (eds.) *Handbook on Organic Crop Production in Sikkim*. Published by Sikkim Organic Mission, Govt. of Sikkim and ICAR RC Sikkim Centre, Sikkim. ISBN N 9 788193 012505 pp. 47–52.

Babu, S., Yadav, G. S., Singh, R., Avasthe, R. K., Das, A., Mohapatra, K. P., Tahashildar, M., Kumar, K., Prabha, M., Devi, M. T., Rana, D. S., Pande, P. and Prakash, N. 2018. Production technology and multifarious uses of buckwheat (*Fagopyrum* spp.): A review. *Ind J Agron* 63(4): 415–427.

Bernath, J. 2000. *Medicinal and Aromatic Plants*. Mezo Publication, Budapest, p. 667.

Bhat, M. R., Bhali, R. K. and Tahir, I. 1986. Predator complex of Melon Aphid (*Aphis gossypii* Glov.), a serious pest of buckwheat (*Fagopyrum* spp.) in Kashmir (India). Buckwheat Newsletter. *Fagopyrum* 6: 12.

Bjorkman, T. 1995. The effect of pollen load and pollen grain competition on fertilization success and progeny performance in *Fagopyrum esculentum. Euphytica* 83(1): 47–52.

Björkman, T. 2012. Northeast Buckwheat Growers Newsletter, No. 33.

Campbell, C. G. and Gubbels, G. H. 1978. *Growing Buckwheat*. Agriculture Canada Publication 1468, Ottawa, Canada.

Cawoy, V., Deblauwe, V., Halbrecq, B., Ledent, J. F., Kinet, J. M. and Jacquemart, A. L. 2006. Morph differences and honey bee morph preference in the distylous species *Fagopyrum esculentum* Moench. *Int J Plant Sci* 167(4): 853–861.

Cawoy, V., Kinet, J. M. and Jacquemart, A. L. 2008. Morphology of nectaries and biology of nectar production in the distylous species *Fagopyrum esculentum* Moench. *Ann Bot* 102(5): 675–684.

Creamer, N. G. and Baldwin, K. R. 1999. *Summer Cover Crops. Horticulture Information Leaflets (HIL-37). North Carolina Cooperative Extension, N.C State University, Raleigh, N.C.* Retrieved November 1, 2011 from www.ces.ncsu.edu/depts/hort/hil/hil-37.html.

Dar, F. A., Pirzadah, T. B., Malik, B., Tahir, I. and Rehman, R. U. 2018. Molecular genetics of buckwheat and its role in crop improvement. In: Zhou, M., Kreft, I., Tang, Y. and Suvorova, G. (eds.) *Buckwheat Germplasm in the World*, 1st edn. Elsevier Publications, USA, pp. 271–286.

Das, A., Babu, H. Subhash, Yadav, G. S., Ansari, M. A., Singh, R., Baishya, R. L. K., Rajkhowa, D. J., and Ngachan, S. V. 2016. Status and strategies for pulses production for food and nutritional security in north-eastern region of India. *Indian J Agron* 61: 43–57.

FAOSTAT 2013. *FAO Statistical Databases*. FAO, Rome. Retrieved from http://faostat.fao.org/ (verified February 24, 2013).

Farooq, S., Rehman, R. U., Pirzadah, R. U., Malik, B., Dar, F. A. and Tahir, I. 2016. Cultivation, agronomic practices and growth performance of buckwheat. In: Zhou, M., Kreft, I., Woo, S. H., Chrungoo, N. and Wieslander, G. (eds.) *Molecular Breeding and Nutritional Aspects of Buckwheat*. Elsevier, USA, pp. 299–319.

Ganyushina, E. V. and Lazarchik, V. M. 1972. Effect of ammonium nitrate on the yield and quality of buckwheat. *Vestn Moskovskogo Univ Biol Pochvoved* 27: 95.

Geneau, C. E., Wackers, F. L., Luka, H., Daniel, C. and Balmer, O. 2011. Selective flowers to enhance biological control of cabbage pests by parastoids. *Basic Appl Ecol* 13: 85–93.

Goncharik, M. N. and Kozlova, A. P. 1972. Effect of chloride ions introduced in potassium fertilizers on the water system of buckwheat leaves. Vyestsi. Akad. Navuk BSSR. *Syer Biyal Navuk* 1: 113–117.

Goodman, R., Hepworth, G., Kaczynski, P., McKee, B., Clarke, S. and Bluett, C. 2001. Honeybee pollination of buckwheat (*Fagopyrum esculentum* Moench) cv. Manor. Aust. *J Exp Agric* 41: 1217–1221.

Halbrecq, B., Romedenne, P. and Ledent, J. F. 2005.Evolution of flowering, ripening and seed set in buckwheat (*Fagopyrum esculentum* Moench): Quantitative analysis. *Eur J Agron* 23(3): 209–224.

Hore, D. and Rathic, R. S. 2002. Collection, cultivation and characterization of buckwheat in Northeastern Region of India. *Fagopyrum* 19: 11–15.

Hulihalli, U. K. and Shantveerayya 2018. Effect of planting geometry and nutrient levels on the productivity of buckwheat. *Int J Curr Microbiol Appl Sci* 7(2): 3369–3374.

Ikanović, J., Rakić, S., Popović, V., Janković, S., Glamočlija, Đ. and Kuzevski, J. 2013. Agro-ecological conditions and morpho-productive properties of buckwheat. *Biotechnol Anim Husb* 29(3): 555–562.

Inamullah, Saqib, G., Ayub, M., Khan, A. A., Anwar, S. and Khan, S. A. 2012. Response of common buckwheat to nitrogen and phosphorus fertilization. *Sarhad J Agric* 28(2): 171–178.

Iqbal, Z., Hiradate, S., Noda, A., Isojima, S. and Fujii, Y. 2003. Allelopathic activity of buckwheat: Isolation and characterization of phenolics. *Weed Sci* 51(5): 657–662.

Jacquemart, A. L., Gillet, C. and Cawoy, V. 2007. Floral visitors and importance of honey bee on buckwheat (*Fagopyrum esculentum* Moench) in central Belgium. *J Hortic Sci Biotechnol* 82(1): 104–108.

Jacquemart, A. L., Cawoy, V., Kinet, J. M., Ledent, J. F. and Quinet, M. 2012. Is buckwheat (*Fagopyrum esculentum* Moench) still a valuable crop today? *Eur J Plant Sci Biotechnol* 6(2): 1–10.

Jung, G. H., Kim, S. L., Kim, M. J., Kim, S. K., Park, J. H., Kim, C. G. and Heu, S. 2015. Effect of sowing time on buckwheat (*Fagopyrum esculentum* Moench) growth and yield in central Korea. *J Crop Sci Biotechnol* 18(4): 285–291.

Kalinova, J., Vrchotova, N. and Triska, J. 2007. Exudation of allelopathic substances in buckwheat (*Fagopyrum esculentum* Moench). *J Agric Food Chem* 55(16): 6453–6459.

Klykov, A. G., Anisimov, M. M., Moiseenko, L. M., Chaikina, E. L. and Parskaya, N. S. 2014. Effect of biologically active substances on morphological characteristics, Rutin content and productivity of *Fagopyrum esculentum* Moench. *Agric Sci Develop* 3(1): 139–142.

Kumar, V., Brainard, D. C. and Bellinder, R. R. 2008. Suppression of Powell amaranth (*Amaranthus powellii*), shepherd's-purse (*Capsella bursa-pastoris*), and corn chamomile (*Anthemis arvensis*) by buckwheat residues: Role of nitrogen and fungal pathogens. *Weed Sci* 56(2): 271–280.

Kuznetsova, A. V. and Klykov, A. G. 2012. Efficiency of chemical and biological preparations in *Rhinoncus sibiricus* Faust control. *J Siberian Messenger Agric Sci* 3: 25–29.

Kuznetsova, A. V., Klykov, A. G., Timoshinov, R. V. and Moiseyenko, L. M. 2012. Harmfulness of *Rhinoncus sibiricus* Faust in Primorsky Krai. *J Rep Russ Acad Agric Sci* 5: 35–38.

Lee, H. B., Kim, S. L. and Park, C. H. 2001. Productivity of whole plant and rutin content under the different quality of light in buckwheat. In: *The Proceeding of the 8th I.S.B*, pp. 84–89.

Loch, J. and Lazanyi, J. 2010. Soil nutrient content in buckwheat production. In: Dobranszki, J (ed.) Buckwheat 2. *The Eur J Plant Sci Biotechnol* 4(1): 93–97.

Ma, M. F., Zheng, S. J., Hiradate, S. and Matsumoto, H. 1997. Detoxifying aluminum with buckwheat. *Nature* 390(6660): 569–570.

Naumkin, V. 2013. Complex of insects on buckwheat plantings. In: *Proceedings of the 12th International Symposium on Buckwheat, Laško*, August 21–25. Fagopyrum, Pernica, pp. 125–126.

Olson, M. 2001. Common buckwheat, agri-facts, agriculture, food and rural management Alberta, Canada. *Open Element*. Retrieved from https://open.alberta.ca/dataset/2551720.

Pavek, P. L. S. 2014. Evaluation of Cover Crops and Plantings Dates for Dryland Eastern Washington Rotations. Plant Materials Technical Note No. 25. United States Department of Agriculture – Natural Resources Conservation Service: Spokane, WA.

Petrich, C. 2000. Phosphorus Mobilization and Weed Suppression by buckwheat. Cropping Systems and Soil Fertility. Minnesota Department of Agriculture. Retrieved November 1, 2011 from www.mda.state.mn.us.

Phogat, B. S. and Sharma, G. D. 2000. Under-utilized food crops: Their uses, adaptation and production technology. In: *Technology Bulletin*, June. NBPGR (ICAR), New Delhi, pp. 10–15.

Pirzadah, T. B., Malik, B., Tahir, I. and Rehman, R. U. 2013. Buckwheat: An introspective and future perspective in Kashmir Himalayas. In: *Proceedings of the 12th International Symposium on Buckwheat, Laško*, August 21–25. Fagopyrum, Pernica, pp. 212–215.

Pirzadah, T. B., Malik, B., Tahir, I. and Rehman, R. U. 2018. Antioxidant potential and ionomic analysis of two buckwheat species from Kashmir region. *Pharmacogn J* 10(6): s83–s88.

Pirzadah, T. B., Malik, B., Tahir, I., Hakeem, K. R., Rehman, R. U. and Alharby, H. F. 2019. Aluminium stress modulates the osmolytes and enzyme defense system in *Fagopyrum* species. *Plant Physiol Biochem* 144: 178–186.

Pirzadah, T. B., Malik, B., Tahir, I. and Rehman, R. U. 2020. Buckwheat journey to functional food sector. *Curr Nutr Food Sci* 16(2): 134–141.

Popović, V., Sikora, V., Berenji, J., Glamočlija, Đ. and Marić, V. 2013a. Effect of agroecological factors on buckwheat yield in conventional and organic cropping systems. *Institute of PKB Agroeconomik: Belgrade* 19(1–2): 155–165.

Popović, V., Sikora, V., Ikanovic, J., Rajičič, V., Maksimović, L. and Katanski, S. 2013b. Production, productivity and quality of buckwheat in organic growing systems in course environmental protection. In: *XVII Eco-Conference, Novi Sad*, September 25–28, pp. 395–404.

Quinet, M., Cawoy, V., Lefevre, I., Van, Miegroet, F., Jacquemart, A. L. and Kinet, J. M. 2004. Inflorescence structure and control of flowering time and duration by light in buckwheat (*Fagopyrum esculentum* Moench). *J Exp Bot* 55(402): 1509–1517.

Rabe, E. and Lovatt, C. J. 1986. Increased arginine biosynthesis during phosphorus deficiency: A response to the increased ammonia content of leaves. *Plant Physiol* 81(3): 774–779.

Racys, J. and Montvilienne, R. 2005. Effect of bee-pollinators in buckwheat (*Fagopyrum esculentum* Moench) crops. *J Apicul Sci* 49: 47–51.

Radics, L. and Mikohazi, D. 2010. Principles of common buckwheat production. In: Dobranszki, J. (ed.) Buckwheat 2. *Eur Plant, J. Sci Biotechnol* 4(1): 57–63.

Rana, J. C., Chauhan, R. C., Sharma, T. R. and Gupta, N. 2012. Analyzing problems and prospects of buckwheat cultivation in India. *Eur J Plant Sci Biotechnol* 6(2): 50–56.

Ratan, P. and Kothiyal, P. 2011. *Fagopyrum esculentum* Moench (common buckwheat) edible plant of Himalayas: A review. *Asian J Pharma Life Sci* 1(4): 426–442.

Ruan, J., Zhou, Y., Yan, J., Zhou, M., Woo, S. H., Weng, W., Cheng, J. and Zhang, K. 2020. Tartary buckwheat: An under-utilized edible and medicinal herb for food and nutritional security. *Food Rev Int.* doi:10.1080/87559129.2020.1734610.

Ruszkowski, M. 1980. The possibility of changing the yielding of buckwheat by breeding homostyle varieties. In: Kreft, I., Javornik, B., and Dolinek, B. (eds.) *Proceedings of the 1st Symp. on Buckwheat, Ljubljana, Yugoslavia.* University of Ljubljana, Biotechnical Faculty, Ljubljana, Slovenia. pp. 7–15.

Small, E. 2017. Buckwheat – The world's most biodiversity-friendly crop? *Biodiversity* 18(2–3): 108–123.

Sobhani, M. R., Rahmikhdoev, G., Mazaheri, D. and Majidian, M. 2014. Influence of different sowing date and planting pattern and N rate on buckwheat yield and its quality. *Aust J Crop Sci* 8(10): 1402–1414.

Sugawara, K. 1956. On buckwheat pollen III. The relation between pollen germination and temperature. *Proc Crop Sci Soc Jpn* 24: 264–265.

Tahir, I. and Farooq, S. 1990. Growth analysis and yield in buckwheats (*Fagopyrum* spp.) grown in Kashmir. *Acta Physiol Plant* 12(4): 311–324.

Tahir, I., Farooq, S. and Bhat, M. R. 1985. Insect pollinators and pests associated with cultivated buckwheat in Kashmir (India). *Fagopyrum* 5: 3–5.

Teboh, J. M. and Franzen, D. W. 2011. Buckwheat (*Fagopyrum esculentum* Moench) potential to contribute solubilized soil phosphorus to subsequent crops. *Commun Soil Sci Plant Anal* 42(13): 1544–1550.

Tsuzuki, E. 2001. Application of buckwheat as a weed control. *Biol Agric Hortic* 76: 55–62.

Tundup, P., Wani, M. A., Hussain, S., Dawa, S. and Tamchos, T. 2017. Traditional methods of buckwheat (*Fagopyrum esculentum* Moench) cultivation in high altitudes cold desert region of India (Ladakh). *Int J Agric Sci Res* 7(1): 101–106.

Wang, C., Ruan, R. W., Yuan, X. H., Hu, D., Yang, H., Li, Y. and Yi, Z. L. 2015. Effects of nitrogen fertilizer and planting density on the lignin synthesis in the culm in relation to lodging resistance of buckwheat. *Plant Prod Sci* 18(2): 218–227.

Warman, P. R. 1991. Effect of incorporated green manure crops on subsequent oat production in an acid, infertile silt loam. *Plant Soil* 134(1): 115–119.

Wirth, J. and Gfeller, A. 2016. Is growing buckwheat allelopathic? *Julius-kühn-Archiv. (Julius Kuhn Inst Bundesforschungsinstitut Kulturpflanz Germany)* 452: 431–438.

Woo, S. H., Kamal, A. H. M., Tatsuro, S., Campbell, C. G., Adachi, T., Yun, S. H., Chung, K. Y. and Choi, J. S. 2010. Buckwheat (*Fagopyrum esculentum* Moench.): Concepts, prospects and potential. *Eur J Plant Sci Biotech* 4: 1–16.

Woo, S. H., Roy, S. K., Kwon, S. J., Cho, S. W., Sarker, K., Lee, M. S., Chung, K. Y. and Kim, H. H. 2016. Concepts, prospects, and potentiality in buckwheat (*Fagopyrum esculentum* Moench): A research perspective. In: Zhou, M., Kreft, I., Woo, S. H., Chrungoo, N. and Wieslander, G. (eds.) *Molecular Breeding and Nutritional Aspects of Buckwheat*. Academic Press, Cambridge, pp. 21–49.

Xuan, T. D. and Tsuzuki, E. 2004. Allelopathic plants: Buckwheat (*Fagopyrum* spp.). *Allelopath J* 13: 137–148.

Zhu, Y. G., He, Y. Q., Smith, S. E. and Smith, F. A. 2002. Buckwheat (*Fagopyrum esculentum* Moench) has high capacity to take up phosphorus (P) from calcium (Ca)-bound source. *Plant Soil* 239(1): 1–8.

3 Journey of Buckwheat to Functional Food Sector

INTRODUCTION

Global food security is dependent on a handful of crops that cannot meet the needs of the growing population, leading to malnutrition and a future food crisis. The decline in the production of crops upon which global food security and economic growth depend has jeopardized future food supplies and rural incomes. There is a dire need to tackle food and feed deficiencies that develop in the form of hidden hunger to attain one of the prime goals of sustainable development programs (Barrett 2010; FAO 2017; Allen and de Brauw 2018). Various stakeholders, including growers, enterprises, research and development organizations and other agencies around the globe, are very eager to revive the cultivation of underexploited and novel crops – so-called neglected/lost crops. These crops were previously used as staple foods, especially in the higher Himalayan regions where the growing period is very short (Wijngaard and Arendt 2006; Pirzadah et al. 2020). While these crops could not compete with the major cereal crops like wheat, maize and rice, they could certainly enhance the food basket to tackle the growing food problems. As per the EASAC policy report, these crops can be strategically explored to assure the efficient management of eco-friendly production of nutraceuticals with minimum inputs and land usage (EASAC 2011). Besides being naturally resilient to various abiotic and biotic stresses, these crops play a pivotal role in mitigating the effects of climate change (Massawe et al. 2015; Mabhaudhi et al. 2019). Buckwheat is regarded as the crop of the 21st century because of its immense potential in the health, food and pharma sectors, and it thus could act as an important source of employment generation for uplifting the socio-economic status of the population (Kwon et al. 2018; Ruan et al. 2020; Sinkovic et al. 2020). However, with the evolution of modern agricultural practices, these traditional crops have fallen into neglect, which has turn led to the genetic erosion of their gene pool. Studies on the cultivation of buckwheat have shown it to be so significant that it has been classified under the category of important crops by FAO due to its declining cultivation and exploitation in nature (Joshi et al. 2019; Pirzadah et al. 2020). Currently, buckwheat has gained great significance in the functional food sector due to its short growing season, climate-resilient nature, rich active metabolites and dietary properties, which make it a promising alternative to staple crops (Ahmed et al. 2013; Alencar and Oliveria 2019). Buckwheat products

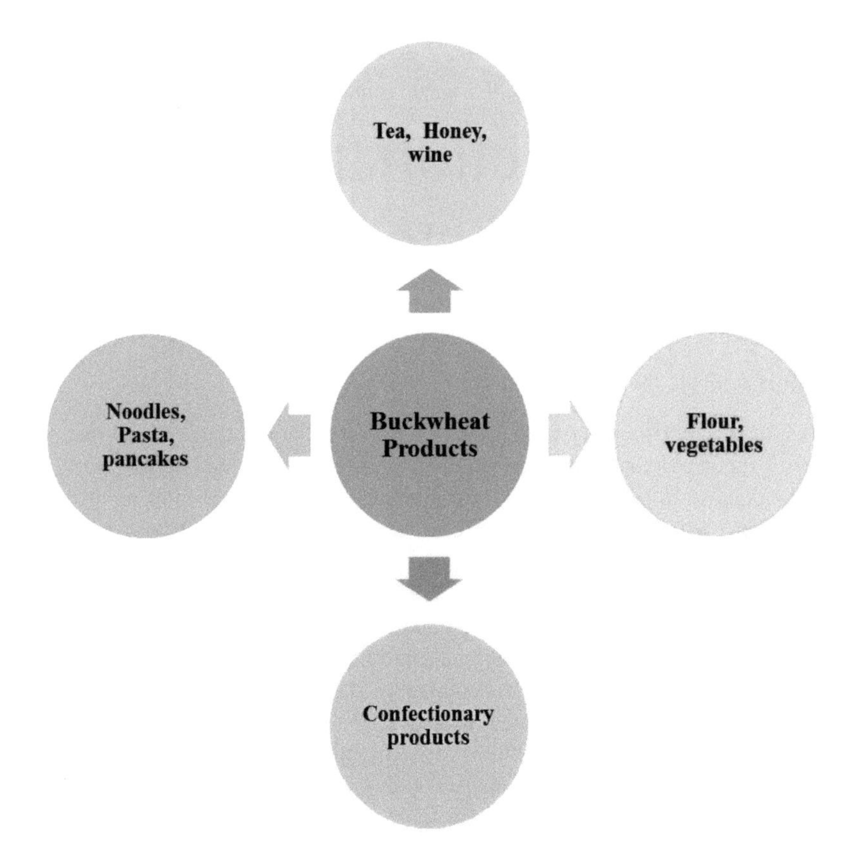

FIGURE 3.1 Commercial products of buckwheat available in the market.

currently available on the market include organic tea, honey, cakes, cookies, biscuits, noodles and wine (Almedia 1978; Qin et al. 2011; Thwe et al. 2013; Pirzadah et al. 2013; Tien et al. 2018; Mohajan et al. 2019) (Figure 3.1). Moreover, due to its gluten-free nature, it has attracted attention particularly from food scientists for the development of gluten-free products for celiac disease patients (Qin et al. 2011). Its predominance in terms of nutritional profile over major grains in terms of the balanced amino acid profile of seed proteins and important metabolites makes it a very unique crop (Table 3.1). In this chapter, we will summarize the importance of buckwheat in terms of nutritional aspects and how its journey proceeds towards the functional food sector.

BUCKWHEAT JOURNEY FROM FIELD TO PLATE

Nutritional Components

Due to the rising demand for gluten-free products and other nutraceuticals, buckwheat has risen rapidly in the functional food sector due to its high nutritional

TABLE 3.1
Nutritional Composition of Buckwheat in Comparison to Three Major Cereals (Data from Campbell 1997; Kumar et al. 2016; Joshi et al. 2019)

Nutrients	Buckwheat	Rice	Wheat	Maize
Proximate composition (per 100 g grain)				
Energy (Kcal)	355	345	346	365
Crude protein (%)	12.0	6.8	11.8	9.4
Total carbohydrates (g)	72.9	78.2	71.2	74.3
Total fiber (%)	17.8	4.5	12.5	7.5
Fat (g)	7.4	1.5	2.5	4.7
Moisture (%)	11.0	11.0	13.7	12.8
Minerals and trace elements (mg/100 g grain)				
Calcium	110	10	30	7
Iron	4.0	0.7	3.5	2.7
Magnesium	390	65	138	127
Phosphorus	330	160	298	210
Manganese	3.4	0.5	2.3	1.9
Zinc	0.8	1.3	2.7	2.3
Potassium	450	268	284	287
Essential amino acids (% of total protein)				
Lysine	5.9	3.8	2.6	1.9
Methionine	3.7	3.0	3.5	3.2
Valine	4.7	5.9	4.5	5.0
Phenylalanine	4.2	5.7	4.4	4.5
Isoleucine	3.5	4.5	3.4	3.8
Threonine	3.5	3.8	2.8	3.9
Histidine	2.2	2.4	2.3	2.4
Cystine	2.2	2.2	1.8	2.2
Tryptophan	1.4	1.0	1.2	0.6
Leucine	6.7	8.2	6.3	13.0
Vitamins (mg/100 g grain)				
Thiamine	3.3	0.06	0.5	0.4
Riboflavin	10.6	0.06	0.2	0.2
Niacin	18.0	1.9	5.5	3.6
Tocopherol	40.0	-	-	-
Pantothenic acid	11.0	-	-	-
Choline	440	-	-	-

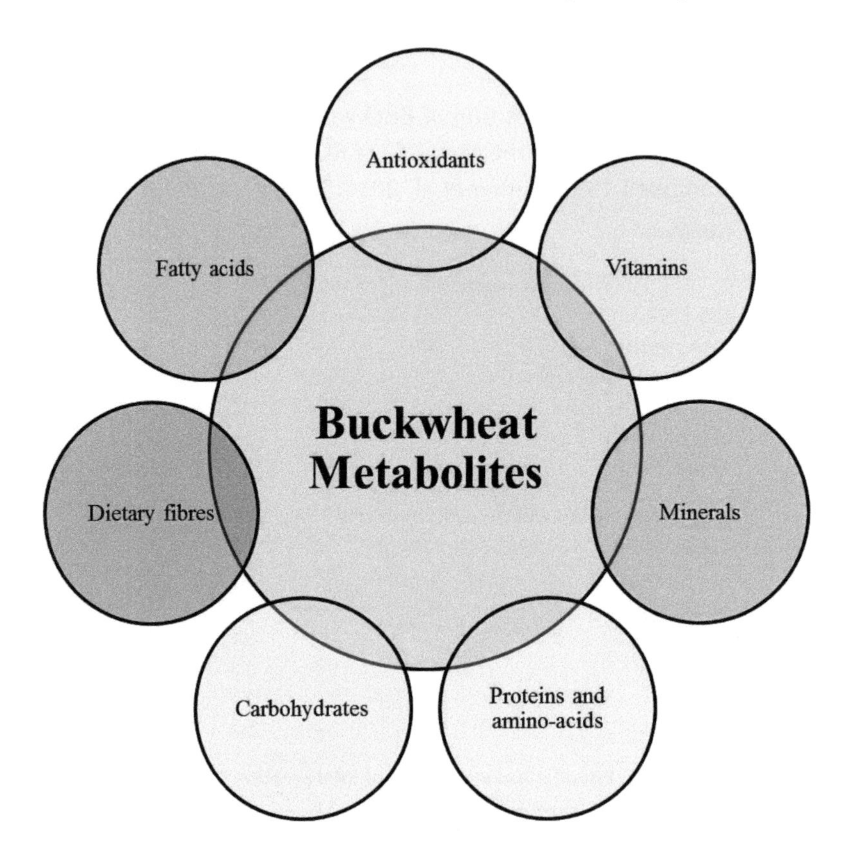

FIGURE 3.2 Various metabolites present in buckwheat.

profile compared to the major staple foods, thus helping sustain the livelihood security of the global population particularly in low-income developing nations where food products based on buckwheat could play a critical role in diversifying nutrition to combat malnutrition problems (Tolaini et al. 2016; Babu et al. 2018; Sytar et al. 2018; Pirzadah et al. 2020). The important metabolites present in buckwheat and its development in the functional food sector will be discussed here (Figure 3.2).

Amino Acids and Proteins

Buckwheat contains high quality proteins with biological values above 90% and well-balanced amino acids compared to major cereal crops; the concentration of proteins usually ranges from 12% to 19% (Ma et al. 1997; Kato et al. 2001; Christa and Soral-Smietana 2008; Zhang et al. 2012; Comino et al. 2013; Gimenez-Bastida et al. 2015; Pirzadah et al. 2018; Kreft et al. 2020). But the concentration of protein is influenced by certain factors that include cultivar type, agronomic practices and eco-physiological factors. Reports also revealed that the protein concentration differs among various parts of the buckwheat plant following the trend bran

> trunk (4%) < embryo (55.9%) > flour (8.5–19%) (Fornal 1999; Krkoškova and Mrazova 2005; Gimenez-Bastida and Zielinski 2015). Based on qualitative analysis, *F. tataricum* exhibits higher protein and amino acid concentration than *F. esculentum* (Wang and Campbell 2004). Further, the protein content of *F. tataricum* is approximately 38.2% greater than rice, 3.9% greater than wheat, 30.5% greater than corn, and 20.2% greater compared to common buckwheat (Wang et al. 1995). The primary constituents of buckwheat proteins include globulins (28.4%), albumins (21.1%), glutelin (36.7%), gliadin (28.4%) and globulin (13.8%) (Ikeda and Asami 2000; Choi et al. 2003; Kumar et al. 2016; Joshi et al. 2019; Pirzadah et al. 2020). Furthermore, the albumin content in buckwheat reaches a maximum level of 30–33% as reported by Bharali and Chrungoo (2003). In common buckwheat, 8S and 13S globulins and 2S albumin are the major storage proteins as reported by Milisavljevic et al. (2004), and these globulins are generally composed of 8S vicilin-like protein types that constitute approximately 12–13 sub-units whose molecular weight varies from 16 to 66 kDa (Krkoškova and Mrazova 2005). However, 13S globulins are salt-soluble proteins and possess a hexameric structure with disulfide linked sub-units that are typically composed of acidic and basic polypeptides with molecular weights varying from 43 to 68, 57 to 58 and 26 to 36 kDa (Radović et al. 1996), while 2S albumins are usually composed of polypeptides with a molecular weight ranging from 8 to 16 kDa (Radović et al. 1999). Due to the presence of higher levels of arginine, aspartic acid and lysine compared to major crops, which are usually devoid of these essential amino acids, buckwheat is considered a promising alternative for super-food applications (Wijngaard and Arendt 2006; Ahmed et al. 2013; Janssen et al. 2016; Sytar et al. 2018). Moreover, due to the low ratio of lysine to arginine ratio (0.79) and methionine to glycine (0.22), particularly in *F. tataricum*, a significant cholesterol-lowering effect was demonstrated (Ge and Wang 2020). The biological value of buckwheat is high due to the presence of a well-balanced amino-acid profile compared to major cereal crops (Skerritt 1986; Sytar et al. 2018). Recently, Sytar et al. (2018) investigated the amino-acid analysis of buckwheat sprouts and concluded that it constitutes high levels of the sulfur-containing amino acid methionine. Moreover, the low level of prolamine and lack of α-gliadin in buckwheat makes it an excellent food source for celiac disease patients (Podolska 2016). The lysine content of buckwheat helps to reduce cholesterol levels by enhancing the fecal excretion of steroids and to suppress gallstone formation by removing bile acids, as well as slowing down the rate of carcinoma of the mammary glands and colorectal cancer (Skrabanja and Kreft 2016; Zhou et al. 2016; Ji et al. 2019; Li 2019; Ge and Wang 2020). Gluten-free buckwheat flour could be used to enhance the digestibility and nutritional value of different buckwheat products by improving their protein quality. Moreover, buckwheat protein's digestibility is lower than cereals because of its high fiber content (17.8%) (Campbell 1997).

Polysaccharides and Dietary Fibers

Buckwheat is made up mainly of carbohydrate, accounting for approximately 75% on a dry weight basis, but this depends on several factors, viz. the type of cultivar, the milling process and other environmental conditions (Qian and Kuhn 1999; Acanski et al. 2015; Zhu 2016; Li 2019). Yilmaz et al. (2020) reported that the carbohydrate content of buckwheat flour is about 70–91% depending upon the efficiency of the extraction process. Li et al. (2001) reported on the starch content of *F. tataricum* (69.84–81.35%) and *F. esculentum* (67.66–86.60%). The amylose content in starch forms the basis for the presence of degenerated starch during the hydrothermal process of food materials (Skrabanja and Kreft 2016). However, the concentration of amylose ranges from 15% to 52% and its degree of polymerization varies from 12 to 45 glucose units (Campbell 1997; Suzuki et al. 2020). From a nutritional point of view, buckwheat starch is categorized into three categories viz., water-resistant starch (2–15μm), slowly digestible starch and rapidly digestible starch; however, their concentration varies depending upon the cooking process (Izydorczyk et al. 2002; Ahmed et al. 2013; Li 2019). Zhu (2016) reported that *F. esculentum* constitutes a more resistant starch than *F. tataricum*. In addition, buckwheat also constitutes about 5–15% analogous carbohydrate, known as dietary fiber, which inhibits absorption and digestion in the small intestine but is partially or fully fermented in the large intestine via microflora. On the basis of nutritional properties, dietary fibers are classified as either soluble or insoluble, and the total dietary fiber concentration in grains is approximately 8.4% compared to soluble and insoluble fiber, which constitutes about 0.2% and 8.2% respectively. Li (2019) reported that dietary fiber content in the groat and seeds is about 7.3% and 10.9% and concluded that these are mainly concentrated in the outer seed covering, which includes the seed coat and hull. Suzuki et al. (2020) conducted a dietary fiber analysis of whole buckwheat flour and reported that the total dietary fiber constituted about 6.7–9.1%, while insoluble and soluble fiber comprised about 2.3–3.2% and 4.3–6.5% respectively. The presence of dietary fiber makes buckwheat an ideal candidate in the pharmaceutical sector to treat various ailments such as diabetes, especially type 2; cardiovascular disease; obesity; hyperglycemia; hypercholesterolemia; and some types of cancer, especially colorectal cancer (Ahmed et al. 2013; Dziedzic et al. 2018; Suzuki et al. 2020).

Iminosugars and D-Chiro-Inositol

Buckwheat is also known to possess iminosugars, and the first iminosugar isolated from the seeds of *F. esculentum* was D-fagomine; its diastereomers, namely 3-epi-fagomine and 3, 4-di-epi-fagomine, were also reported in the bran, leaves, flour and groats (Amezqueta et al. 2012). Joshi et al. (2020) reported that *F. esculentum* possesses a higher content of D-fagomine than *F. tataricum*, and the concentration of 3, 4-di-epi-fagomine (43mg/kg) and D-fagomine (44mg/kg) was found to be more in the groats. Due to the high bioactivities of these iminosugars, they are found to be excellent glycosidase inhibitors and thus find great application in the functional food sector. It has been reported that D-fagomine possesses immense medicinal properties such as lowering the chances of increasing insulin resistance, sucrose-induced hypertension, F2-IsoPs (oxidative stress markers); it also has an anti-hyperglycaemic effect and can reduce an excess of potentially pathogenic bacteria, steatosis, uric acid and liver diacylglycerols (DAGs), without broadly affecting perigonadal fat deposition and impaired glucose tolerance (Dziedzic et al. 2018; Joshi et al. 2020; Ramos-Romero et al. 2020).

Reports revealed that D-chiro-inositol, which is found in abundant quantities especially in Farinetta (a type of buckwheat flour), is an insulin mediator that acts as a secondary messenger with a vital role in insulin signal transduction pathways, which is usually evident in patients with type II diabetes and polycystic ovary syndrome (Li et al. 2019; Martin-Garcia et al. 2019; Ramos-Romero et al. 2020). The important derivative of D-chiro-inositol present in buckwheat is fagopyritols, which are primarily concentrated in the aleurone layer and the buckwheat seed embryo (Horbowicz et al. 1998). Wu et al. (2018) reported that *F. tataricum* constitutes about 11 times more D-chiro-inositol. Buckwheat is known to possess six types of fagopyritols, viz. A1, A2, A3, B1, B2, and B3, of which A1 and B1 are the principal fagopyritols (Zhu 2016; Wu et al. 2018; Suzuki et al. 2020). Moreover, due to the low glycemic index (GI=34.7) of buckwheat, it helps to maintain blood sugar levels (Rozanska et al. 2020).

Fatty Acid Content

Buckwheat seeds also constitute monosaturated as well as polyunsaturated fatty acid that differs among different cultivars and plays an essential role in determining the quality of food (Campbell 1997; Subedi 2018; Ge and Wang 2020). *F. tataricum* and *F. esculentum* have a higher fatty acid content (3.9%) compared to cereal crops, and the main components include oleic (47%), linoleic (39%), linolenic (4%), stearic (2.5%) and palmitic acid (5.8%) (Horbowicz and Obendorf 1992; Campbell 1997; Pirzadah et al. 2017; Li 2019; Ge and Wang 2020). It has been reported that lipids are mainly concentrated in the buckwheat embryo (7–14%) and this is thus regarded as the richest source of lipids (Bonafaccia et al. 2003; Ahmed et al. 2018). The lipid concentration among various seed parts follows the order hull (0.4–0.7%) < endosperm (2–3%) < embryo (9.7–19.7%) (Subedi 2018). The total lipid concentration of buckwheat is comparable to that of wheat rye but it is about 72.4% higher than rice and 48.2% more than wheat (Becker 2008). Moreover, long-chain fatty acids, which include arachidic, behenic and lignoceric acids, constitute approximately 80mg/g of the total acids in buckwheat, and are the only minor metabolites that are usually lacking in major cereals. Subedi (2018) reported that the lipid concentration of *F. tataricum* and *F. esculentum* ranges from 1.2% to 4.3% and 1.5% to 4.0% respectively. Recently, fatty acid analysis was done on *F. tataricum* and it was concluded that 84% of the fatty acid content is concentrated in the bran, as compared to about 83% in the flour (Ge and Wang 2020). In addition, buckwheat also comprises unsaturated fatty acids (74.79%), primarily concentrated in the embryo, and saturated fatty acids (25%), which are found mainly in the hull. The seed coat of buckwheat is a rich source of an essential fatty acid, viz., linoleic acid, with respect to other organs. Due to the presence of polyunsaturated fatty acids, buckwheat plays a pivotal role in lowering the likelihood of myocardial infarction (Ge and Wang 2020).

Macro- and Micro-Nutrients

Buckwheat is also a rich source of minerals, especially the trace elements P, Cu, B, I, Fe, Se, Ba, Zn and Co, which are primarily concentrated in the aleurone tissue, embryo and hull, as reported by Subedi (2018). Moreover, it also constitutes macro-nutrients like Ca, K, Na and Mg (Stibilj et al. 2004; Pirzadah et al. 2013; Pirzadah et al. 2017). However, the concentration of these minerals varies among different parts and follows the order flour (0.8–9%) > hull (3.4–4.2%) > dehulled grains (2.2–3.5%) > whole grains (2.0–2.5%) > kernel (1.8–2.0%) (Li and Zhang 2001; Christa and Soral-Smietana 2008). The difference in the mineral content may be due to agro-climatic or genetic factors or post-harvesting techniques. Previous reports also reported that *F. tataricum* has more Fe, Se, Ni, Zn and Co content than *F. esculentum* (Zhu 2016). Ikeda et al. (1999) investigated the mineral analysis of buckwheat and other major cereals and concluded that buckwheat possesses higher levels of Zn, Mn and Cu than maize, wheat or rice. Ikeda et al. (2006) performed a comparative analysis between buckwheat and other cereal flours using an *in vitro* enzymatic digestion technique and observed that buckwheat flour constitutes greater levels of essential minerals compared to other cereal flour. In addition, it has also been proven that a majority of the Zn, Cu and K of buckwheat flour is generally released in soluble form compared to cereal flour. It has been reported that 100g of buckwheat flour could provide about 13–89% of the recommended dietary allowance (RDA) for Zn, Cu, Mg and Mn but just a few percent of the RDA for Ca (Pirzadah et al. 2020).

Vitamins

Other important constituents that make buckwheat an efficient nutraceutical as compared to cereals include vitamins, especially vitamin B_1 (thiamine), B_2 (riboflavin), B_3 (niacin and niacinamide), vitamin B_5 (pantothenic acid), vitamin B_6 (pyridoxine), vitamin C, (ascorbic acid) and vitamin E (tocopherol) (Alvarez-Jubete et al. 2010; Subedi 2018). However, the concentration of vitamins like B1 and C is enhanced upon germination as reported by Zhou et al. (2015). Vitamin E in buckwheat exists in different forms in the seeds, such as γ-tocopherol (117.8 μg/g), δ-tocopherol (7.3 μg/g) and α-tocopherol (2.1 μg/g) (Pirzadah et al. 2017; Li 2019). It has also been reported that *F. tataricum* possesses a higher content of B-group vitamins than *F. esculentum*, which is rich in vitamin E, and these are mainly concentrated in the peripheral regions of the endosperm and embryo (Ikeda et al. 2006; Joshi et al. 2020). However, the concentrations of these vitamins are altered by various factors like developmental stage and agro-climatic factors. Kim et al. (2002) reported that the achene of buckwheat contains vitamin B in the order niacin (B_3) 18.0 mg/kg > pantothenic acid (B_5) 11.0 mg/kg > riboflavin (B_2) 10.6 mg/kg > thiamin (B_1) 3.3 mg/kg > pyridoxine (B_6) 1.5 mg/kg. In another study, Kim et al. (2004) carried out an experimental analysis and concluded that vitamin C increases up to 0.25mg/g in the sprouts of buckwheat, while the total tocopherol content ranges from 14.3 to 21.7 mg/kg.

METABOLOMICS DISSECTION OF BUCKWHEAT

Buckwheat is a treasure trove of various metabolites, such as phenolic and flavonoid compounds, which occur in various parts like the seeds, sprouts, groat, hull, flowers, leaves and roots, as usually detected by various metabolomic techniques (GC-MS and LC-MS) (Pirzadah et al. 2017; Park et al. 2017; Matsui and Walker 2019; Borovaya and Klykov 2020) (Table 3.2). However, the concentrations of these metabolites vary depending upon numerous factors like the agro-climatic conditions, type of cultivar, sowing season and other post-harvesting technologies, as reported by various authors (Matsui and Walker 2019). *F. tataricum* usually exhibited more flavonoid content, approximately 40mg/g more than *F. esculentum*, which constitutes about 10mg/g as reported by Li (2019). It has also been revealed that the stem, leaves and flowers of *F. tataricum* constitute about 100mg/g flavonoid content as compared to the leaves and flowers of *F. esculentum*, which possess about 1.2–2.6% and 8.3–10% respectively (Li 2019; Borovaya and Klykov 2020). The main polyphenolic metabolites present in the buckwheat are as follows:

RUTIN

Rutin is the major flavonol present in various parts of buckwheat species (flowers, sprouts, leaves, groat and hull) and constitutes about 90% of the total phenolic compounds as reported by various authors (Jiang et al. 2007; Jing et al. 2016; Pirzadah et al. 2017; Sytar et al. 2018). However, the concentration of rutin varies among different species of buckwheat, viz. *F. tataricum* has a higher rutin content than *F. esculentum* (Kitabayashi et al. 1995; Park et al. 2011; Zhu 2016; Borovaya and Klykov 2020). In *F. esculentum*, the rutin content follows the order roots (3–8mg/g) < stem (6–14mg/g) < flowers (47–63mg/g), as reported by Borovaya and Klykov (2020). In addition, the rutin content also varies depending upon numerous factors like growth conditions, developmental stage, plant part, and harvesting time (Ohsawa and Tsutsumi 1995). It has been reported that the maximum amount of rutin was found in flowers and leaves during the seed development and flowering stage. Other factors that affect the rutin content involve the processing technique and endogenous enzymes such as 3-o-β-heterodiasaccharidase (FHG), often referred to as rhamnodiastase or rutinosidase, which causes the degradation of rutin by hydrolyzing the glycosidic bond of rutin (Baumgertel et al. 2003; Taguchi 2016). In addition, enzymes like polyphenol oxidase (PPO) could also contribute to the breakdown of rutin and quercetin. Thus, the appropriate time for harvesting buckwheat is in the early morning hours during early flowering, because this is when the rutin content of flowers is high and the rutin degrading enzyme activity is low (Ohsawa and Tsutsumi 1995; Baumgertel et al. 2003; Pirzadah et al. 2020). The physiological function of rutin involves the protection of buckwheat from various abiotic and biotic stresses such as cold, drought, UV rays and insects. Moreover, a high rutin concentration alters the taste of buckwheat (its bitterness) and this property helps to prevent it from

TABLE 3.2
Summary of the Various Metabolites from Buckwheat Using Different Metabolomics Methods

Category	Compounds	Detection Methods	References
Flavonoids	Isoquercetin, quercetin, and rutin	HPLC	Gabr et al. (2019); Ge and Wang (2020), Kalinova et al. (2019)
	Rutin	GC-MS	Pirzadah et al. (2017)
		HPLC-ESI-MS and HPLC-UV	Park et al. (2019)
		UPLC-ESI-MS/MS	Li et al. (2019)
		HPLC-MS	Martin-Garcia et al. (2019)
	Hyperoside and quercetin	RP-UHPLC-ESI-MS	Dziedzic et al. (2018)
		UPLC-ESI-MS/MS	Li et al. (2019)
		HPLC	Kalinova et al. (2019)
	Procyanidin B2	RP-UHPLC-ESI-MS	Dziedzic et al. (2018)
		HPLC-MS	Martin-Garcia et al. (2019)
		HPLC	Kalinova et al. (2019)
	Luteolin	RP-UHPLC-ESI-MS	Dziedzic et al. (2018)
	Kaempferol	RP-UHPLC-ESI-MS	Dziedzic et al. (2018); Pirzadah et al. 2017
		UPLC-ESI-MS/MS	Li et al. (2019)
		HPLC	Gabr et al. (2019)
	Catechin	RP-UHPLC-ESI-MS	Dziedzic et al. (2018)
	Catechin and epicatechin	HPLC-PAD and LIT-FTICR-MS	Zhu (2016)
	Catechin, epicatechin, and epiafzelchin	HPLC-MS	Martin-Garcia et al. (2019)
	Catechin, epicatechin, and epicatechin gallate	HPLC	Kalinova et al. (2019)
	Fagopyrin A to fagopyrin F	RP-UHPLC-ESI-MS and NMRS	Joshi et al. (2020)
	Isovitexin and vitexin	RP-UHPLCESI-MS	Dziedzic et al. (2018)
		UPLC-ESI-MS/MS	Li et al. (2019a)
		HPLC	Kalinova et al. (2019)
	Isoorientin, isovitexin, orientin and vitexin	HPLC	Nam et al. (2018); Kalinova et al. (2019)
		HPLC-ESI-MS, HPLC-UV	Park et al. (2019)
		UPLC-ESI-MS/MS	Li et al. (2019a)
		HPLC-MS	Martin-Garcia et al. (2019)

(*Continued*)

TABLE 3.2 (CONTINUED)
Summary of the Various Metabolites from Buckwheat Using Different Metabolomics Methods

Category	Compounds	Detection Methods	References
Anthocyanins	Cyanidin, cyanidin-3-*O*-glucoside, cyanidin-*O*-syringic acid, cyanidin-3-*O*-glucosylmalonylglucoside, peonidin, petunidin-3-*O*-glucoside and cyanidin-3-*O*-rutinocide	UPLC-ESI-MS/MS	Li et al. (2019a)
	Cyanidin-3-*O*-glucoside and cyanidin-3-*O*-rutinocide	HPLC-ESI-MS	Zhu (2016)
Anthraquinones	Aloe-emodin, aurantio-obtusin, chryso-phanol, emodin, rhein and physcion	HPLC-DAD and UPLC-DAD	Zhu (2016)
Phenolic acids	Caffeic acid, chlorogenic acid, ferulic acid, gallic acid, 4-hydrobenzoic acid, isovanilic acid, *p*-coumaric acid,	RP-UHPLC-ESI-MS	Dziedzic et al. (2018)
	p-hydroxybenzoic, and syringic acids, Caffeic acid, chlorogenic acid, ferulic acid, gallic acid, *p*-coumaric acid, *p*-hydroxybenzoic, protocatechuic, syringic and vanillic acids	HPLC, MS and NMRS	Zhu (2016)
	Vanillic acid, vanillin and protocatechuic acid	HPLC	Kalinova et al. (2019)
Stilbene	Resveratrol	HPLC	Zhu (2016)
Fagopyrins	Fagopyrin A–F	NMRS and MS	Joshi et al. (2020)
		HPLC-UV-vis photometry	Zhu (2016)
Fagopyritol	Fagopyritol A1 and B1	GC-MS and NMRS	Wu et al. (2018)
Steroids	β-sitosterol, β-sitosterol palmitate, daucosterol, ergosterol peroxide, stigmsat-4-en-3,6-dione, stigmast-5-en-3-ol	Capillary GC/MS	Zhu (2016)
	β-sitosterol, β-sitosterol palmitate, stigmast-5-en-3-ol	GC-MS	Pirzadah et al. 2017
Triterpenoids	Glutinone, glutinol, olean-12-en-3-ol and urs-12-an-3-ol	Capillary GC/MS	Jing et al. (2016)

(Continued)

TABLE 3.2 (CONTINUED)
Summary of the Various Metabolites from Buckwheat Using Different Metabolomics Methods

Category	Compounds	Detection Methods	References
Phenylpropanoid glycosides	Diboside A and tatarisides A–G	HPLC-PDA/ LTQ-FTICR-MS, NMRS and MS	Zhu (2016)
Proteins	Amino acid compositions	HPLC	Tien et al. (2018)
Carbohydrates	Polysaccharides and monosaccharides	IRS, GC, GC-MS, NMRS and HPLC	Ji et al. (2019)
	D-*chiro*-inositol	HPLC- ELSDs	Zhu (2016)
Fatty acids	Fatty acid compositions	GC	Sinkovic et al. (2020)
	Free fatty acid compositions	GLC	Tien et al. (2018)
Vitamins	Vitamins B1, B6 and C	HPLC	Kim et al. (2004)
Carotenoids	Lutein and β-carotene	HPLC-UV-HG-AFS	Tuan et al. (2013)

Abbreviations: *HPLC* – High-performance liquid chromatography; *HPLC-ESI-MS* – High-performance liquid chromatography-electrospray ionization-mass spectrometry; *HPLC-UV* – High-performance liquid chromatography-UV analyses; *UPLC-ESI-MS/MS* – Ultra performance liquid chromatography–electrospray ionization–tandem mass spectrometry system; *HPLC-MS* – High-performance liquid chromatography-mass spectrometry; *RP-UHPLC-ESI-MS* – Reverse-phase ultra-performance liquid chromatography electrospray ionization-mass spectrometry; *HPLC-PAD* – High performance liquid chromatography with photo-diode array detector; *LIT-FTICR-MS* – Linear ion trap Fourier transform ion cyclotron resonance hybrid mass spectrometry; *NMRS* – Nuclear magnetic resonance spectroscopy; *HPLC-DAD* – High-performance liquid chromatography with diode array detector; *UPLC–DAD* – Ultra performance liquid chromatography with diode array detector; *MS* – Mass spectrometry; *GC-MS* – Gas chromatography-mass spectrometry; *HPLC-PDA/LTQ-FTICR-MS* – High performance liquid chromatography photo-diode array detector/linear ion trap Fourier transform ion cyclotron resonance hybrid mass spectrometry; *IRS* – Infrared spectroscopy; *GC* – Gas chromatography; *HPLC-ELSDs* – High-performance liquid chromatography-evaporative light-scattering detectors; *GLC* – Gas-liquid chromatography; *HPLC-UV-HG-AFS* – High performance liquid chromatography-UV irradiation-hydride generation-atomic fluorescence spectrometry.

being grazed by animals (Kreft et al. 2020; Joshi et al. 2020). Metabolomic analysis of buckwheat has revealed some other important metabolites such as vitexin, isovitexin, orientin, fagopyrins and homoorientin (Zhu 2016; Kwon et al. 2018; Matsui and Walker 2019).

Nutraceutical Benefits of Rutin

Rutin plays a prominent role in the nutraceutical sector because of its many medicinal properties, such as being anti-oxidant, anti-cancer, anti-hypertensive, spasmolytic and vasoconstructive (Landberg et al. 2011; Sharmila et al. 2014).

Moreover, it enhances the elasticity of arteries and capillaries thus making them strong and flexible. In addition, previous studies have also reported that rutin helps to protect the skin from UV-rays, inhibit gastric lesions and improve vision and hearing, as well as inhibit the formation of gallstones (Gong et al. 2010; Kuntic et al. 2011; Gimenez-Bastida and Zielinski 2015). It has been reported that adding rutin to flour aids in pepsin digestibility (Guo et al. 2007). Furthermore, rutin is also involved in reducing the rate of protein synthesis and myocyte cell surface area, thus preventing the proliferation of fibroblasts and showed strong inhibition on cardiac hypertrophy. Besides, it also restricts the cardiac hypertrophy induced by angiotensin II (Han et al. 2010). Rutin reduces glycolipotoxic effects by activating cAMP-dependent protein kinase signaling, leading to an inhibition of the activities of lipogenic enzymes, which in turn enhances mitochondrial functions and thus could be used to treat type II diabetes (Cai and Lin 2009). Awatsuhara et al. (2010) reported the strong antioxidant activity of rutin to scavenge $OH^{\bullet}$ radicals, thus helping to prevent DNA damage. In another study, rutin was found to act as antagonistic to some drugs (salicylates, sulphadiazine and thiocyanates) that are known to weaken the capillaries (Clemetson 1976). In addition, rutin has also been shown to be antagonistic to ascorbic acid, the conversion of which to dehydroascorbic acid can contribute to hypertension, diabetes mellitus and cardiovascular diseases (Clemetson 1976). Currently, the known herbal drug Fagopyri herba, which has rutin as its main component, is used to treat cardiovascular problems (Ihme et al. 1996).

The major polyphenolic compounds found during metabolomic dissection in various buckwheat species are summarized in Table 3.3.

HEALTH CLAIMS FOR BUCKWHEAT

Buckwheat is an ancient staple crop with a high nutritional profile that is consumed worldwide (Pirzadah et al. 2013; Kwon, et al. 2018; Park et al. 2019; Sinkovic et al. 2020). Chinese quotes include "People who love buckwheat are healthy" and "People who love buckwheat live long." In recent years, demand for buckwheat has increased and caught the attention of food scientists because of its identical chemical compounds and the fact that it is very effective as a nutraceutical food with curative effects on chronic diseases; it has anti-hypertensive, anti-diabetic, anti-oxidative, cardioprotective, anti-cancer, hepatoprotective, anti-tumor, anti-inflammatory, cholesterol-lowering, neuroprotective and cognitive-enhancing effects (Ihme et al. 1996; Lv et al. 2017; Kwon et al. 2018; Ge and Wang 2020) (Figure 3.3, Table 3.4). Due to its enrichment with abundant nectar, buckwheat is a significant source of honey. In Korea, various products are made with buckwheat flour, including Naengmyeon, Makguksu and Mug; Makguksu in particular is prepared according to scientific instructions from the blend of *Eucommia ulmoides*, bitter buckwheat and other medicinal herbs and has potential nutraceutical properties due to its easy absorption into the blood. Moreover, due to the gluten-free nature of buckwheat, it is used as a promising alternative for celiac disease patients

TABLE 3.3
Summary of the Metabolomics Profiling of Different Buckwheat Species (Jing et al. 2016; Lv et al. 2017; Pirzadah et al. 2017; Dziedzic et al. 2018; Gorniak et al. 2019; Martin-Garcia et al. 2019; Park et al. 2019; Melini et al. 2020)

Compound Name	Example	Source
	Flavonoid	
Flavonols		
	Rutin	*Fc, Fe, Fh, Ft*
	Kaempferol	*Fc, Ft*
	Kaempferol-3-*O*-galactoside	*Ft*
	Kaempferol-3-*O*-glucoside	*Ft*
	Kaempferol-3-*O*-rutinoside	*Ft*
	Kaempferol-3-*O*-sophoroside	*Fe*
	Kaempferol-3-*O*-glucoside-7-*O*-glucoside	*Fe*
	Myricetin	*Fe*
	Quercetin	*Fc, Fe, Ft*
	Isoquercetin	*Fc, Fe*
	Quercitrin (quercetin-3-*O*-rhamnoside)	*Fc, Ft*
	Isoquercitrin	*Fe*
	Quercetin-3-*O*-[β-D-xyloxyl-(1-2)-α-L-rhamnoside]	*Ft*
	Quercetin-3-*O*-β-D-galactoside	*Fe, Ft*
	Quercetin-3-*O*-rutinoside-3'-*O*-β-glucopyranoside	*Fc, Ft*
	Quercetin-3-*O*-rutinoside-7-*O*-galactoside	*Ft*
	Rhamnetin	*Fc*
Flavones		
	Luteolin	*Fc*
	Vitexin	*Fe, Ft*
	Isovitexin	*Fe, Ft*
	Orientin	*Fe, Ft*
	Isoorientin	*Fe, Ft*
	Homoorientin	*Fe*
	Quercetin-3-*O*-rutinoside-3'-*O*-glucoside	*Ft*
	Quercetin-3-*O*-rutinoside-7-*O*-galactoside	*Ft*
	3',4'-methylenedioxy-7-hydroxy-6-isopentenyl flavone	*Fc*
Flavanones		
	Hesperetin 7-rutinoside (hesperidin)	*Fe, Ft*
	Hesperetin 7-*O*-neohesperidoside	*Fe, Ft*
	Hesperetin *O*-hexosyl-*O*-hexoside	*Fe, Ft*
	Hesperetin 5-*O*-glucoside	*Fe, Ft*
	Hesperetin *O*-malonylhexoside	*Fe, Ft*
	Naringenin	*Fe, Ft*

(Continued)

TABLE 3.3 (CONTINUED)
Summary of the Metabolomics Profiling of Different Buckwheat Species (Jing et al. 2016; Lv et al. 2017; Pirzadah et al. 2017; Dziedzic et al. 2018; Gorniak et al. 2019; Martin-Garcia et al. 2019; Park et al. 2019; Melini et al. 2020)

Compound Name	Example	Source
	Naringeninchalcone	*Fe, Ft*
	Naringenin *O*-malonylhexoside	*Fe, Ft*
	Naringenin 7-*O*-glucoside	*Fe, Ft*
	Phloretin	*Fe, Ft*
	Homoeriodictyol	*Fe, Ft*
	Hesperetin	*Fc*
	(-)-Liquiritigenin	*Ft*
Flavanols/Flavan-3-ols		
	Catechin	*Fc, Fe, Fh*
	(+)-catechin-7-*O*-glucoside	*Fe*
	Catechin hydrate	*Ft*
	Epicatechin	*Fc, Fe, Fh, Ft*
	Epicatechin-3-*O*-(3,4-di-*O*-methyl)-gallate	*Fe, Fh*
	(–)-epicatechin-3-*O*-*p*-hydroxybenzoate	*Fe*
	Epigallocatechin	*Ft*
	Epicatechingallate	*Fe, Fh*
	Epiafzelechin-(4–6)-epicatechin	*Fe, Fh*
	Epiafzelechin-(4–8)-epicatechin-*p*-OH-benzoate	*Fe, Fh*
	Epiafzelechin-(4–8)-epicatechin-methylgallate	*Fe, Fh*
	Epicatechin(4–8)-epicatechin-*O*-(3,4-dimethyl)-gallate	*Fe, Fh*
	Epiafzelechin-(4–8)-epicatechin(3,4-dimethyl)-gallate	*Fe, Fh*
	Epiafzelechin-(4–8)-epiafzelechin-(4–8)-epicatechin	*Fe*
	Epiafzelechin-(4–8)-epiafzelechin-(4–8)-epicatechin-*O*-(3,4-dimethyl)-gallate	*Fe, Fh*
Anthocyanins		
	Cyanidin 3-*O*-glucoside	*Fe, Ft*
	Cyanidin 3-*O*-rutinoside	*Fe, Ft*
	Cyanidin 3-*O*-galactoside	*Fe*
	Cyanidin 3-*O*-galactopyranosyl-rhamnoside	*Fe*
Fagopyrins		
	Fagopyrin A	*Fc, Fe, Ft*
	Fagopyrin B	*Fc, Fe, Ft*
	Fagopyrin C	*Fc, Fe, Ft*
	Fagopyrin D	*Fc, Fe, Ft*
	Fagopyrin E	*Fc, Fe, Ft*
	Fagopyrin F	*Fc, Fe, Ft*

(*Continued*)

TABLE 3.3 (CONTINUED)
Summary of the Metabolomics Profiling of Different Buckwheat Species (Jing et al. 2016; Lv et al. 2017; Pirzadah et al. 2017; Dziedzic et al. 2018; Gorniak et al. 2019; Martin-Garcia et al. 2019; Park et al. 2019; Melini et al. 2020)

Compound Name	Example	Source
Proanthocyanidins		
	Procyanidin A1	
	Procyanidin A2	*Fe, Fh*
	Procyanidin A3	*Fe, Ft*
	Procyanidin B2	*Fe, Ft*
	Procyanidin B3	*Fe, Ft*
	Procyanidin B5	*Fe, Ft*
	Procyanidin C1	*Fe, Fh*
Isoflavones		
	6-hydroxydaidzein	*Fc*
	2'-hydroxydaidzein	*Fe, Ft*
	Sissotrin	*Fe, Ft*
	Formononetin (4'-*O*-methyldaidzein)	*Fe, Ft*
	Glycitin	*Fe, Ft*
	Genistein 7-*O*-glucoside (genistin)	*Fe, Ft*
	Formononetin 7-*O*-glucoside (Ononin)	*Fe, Ft*
Flavonolignan		
	Tricin 4'-*O*-(β-guaiacylglyceryl) ether *O*-hexoside	*Fe, Ft*
	Tricin 7-*O*-β-guaiacylglycerol	*Fe, Ft*
	Tricin 4'-*O*-β-guaiacylglycerol	*Fe, Ft*
	Tricin 4'-*O*-syringic acid	*Fe, Ft*
	Tricin 4'-*O*-(syringyl alcohol) ether 5-*O*-hexoside	*Fe, Ft*
	Tricin 4'-*O*-(syringyl alcohol) ether 7-*O*-hexoside	*Fe, Ft*
	Phenolic acids and their derivatives	
Hydroxybenzoic acids		
	Benzoic acid	*Fc*
	Gallic acid	*Fc, Ft*
	4-hydroxybenzoic acid	*Ft*
	P-hydroxybenzoic acid	*Fc*
	P-hydroxybenzaldehyde	*Ft*
	Vanillic acid	*Ft*
	Protocatechuic acid	*Fc, Fe, Ft*
	Syringic acid	*Ft*
Hydroxycinnamic acids		
	P-coumaric acid	*Ft*
	O-coumaric acid	*Ft*

(*Continued*)

TABLE 3.3 (CONTINUED)
Summary of the Metabolomics Profiling of Different Buckwheat Species (Jing et al. 2016; Lv et al. 2017; Pirzadah et al. 2017; Dziedzic et al. 2018; Gorniak et al. 2019; Martin-Garcia et al. 2019; Park et al. 2019; Melini et al. 2020)

Compound Name	Example	Source
	Caffeic acid	*Ft*
	Ferulic acid	*Ft*
	2,4-dihyroxycinnamic acid	*Ft*
	Chlorogenic acid	*Fe*, *Ft*
Stilbenes		
	Resveratrol	*Fe*, *Ft*
Steroids		
	Hecogenin	*Fc*
	β-sitosterol	*Fc*, *Ft*
	β-sitosterol palmitate	*Ft*
	Ergosterol peroxide	*Ft*
	Daucosterol	*Ft*
	β-daucosterol	*Fc*
	6-hydroxy stigmasta-4,22-dien-3-one	*Fe*
	23S-methylcholesterol	*Fe*
	Stigmast-5-en-3-ol	*Fe*
	Stigmast-5,24-dien-3-ol	*Fe*
	Trans-stigmast-5,22-dien-3-ol	*Fe*
	Stigmast-4-ene-3,6-dione	*Ft*
Triterpenoids		
	Ursolic acid	*Fc*, *Ft*
	Olean-12-en-3-ol	*Fe*
	Urs-12-en-3-ol	*Fe*
	α-thujene	*Ft*
	α-terpineol	*Ft*
	Glutinone	*Fc*
	Glutinol	*Fc*
Tannins		
	3,3-di-*O*-galloyl-procyanidin B-2	*Fc*
	3-*O*-galloyl-procyanidin B-2	*Fc*
Phenylpropanoid glycosides		
	Tatarisides A	*Ft*
	Tatarisides B	*Ft*
	Tatarisides C	*Ft*
	Tatarisides D	*Ft*
	Tatarisides E	*Ft*

(*Continued*)

TABLE 3.3 (CONTINUED)
Summary of the Metabolomics Profiling of Different Buckwheat Species (Jing et al. 2016; Lv et al. 2017; Pirzadah et al. 2017; Dziedzic et al. 2018; Gorniak et al. 2019; Martin-Garcia et al. 2019; Park et al. 2019; Melini et al. 2020)

Compound Name	Example	Source
	Tatarisides F	*Ft*
	Tatarisides G	*Ft*
	Diboside A	*Fc, Ft*
	Lapathoside A	*Fc*
	3,6-di-*p*-coumaroyl-1,6'-di-feruloyl sucrose	*Ft*
	1,3,6'-tri-feruloyl-6-*p*-coumaroyl sucrose	*Ft*
	1,3,6-tri-*p*-coumaroyl-6'-feruloyl sucrose	*Ft*
	1,3,6,6'-tetra-feruloyl sucrose	*Ft*

Abbreviations: *Fe*, *Ft*, *Fc*, *Fh* denote *Fagopyrum esculentum*, *F. tataricum*, *F. cymosum* and *F. homotropicum* respectively.

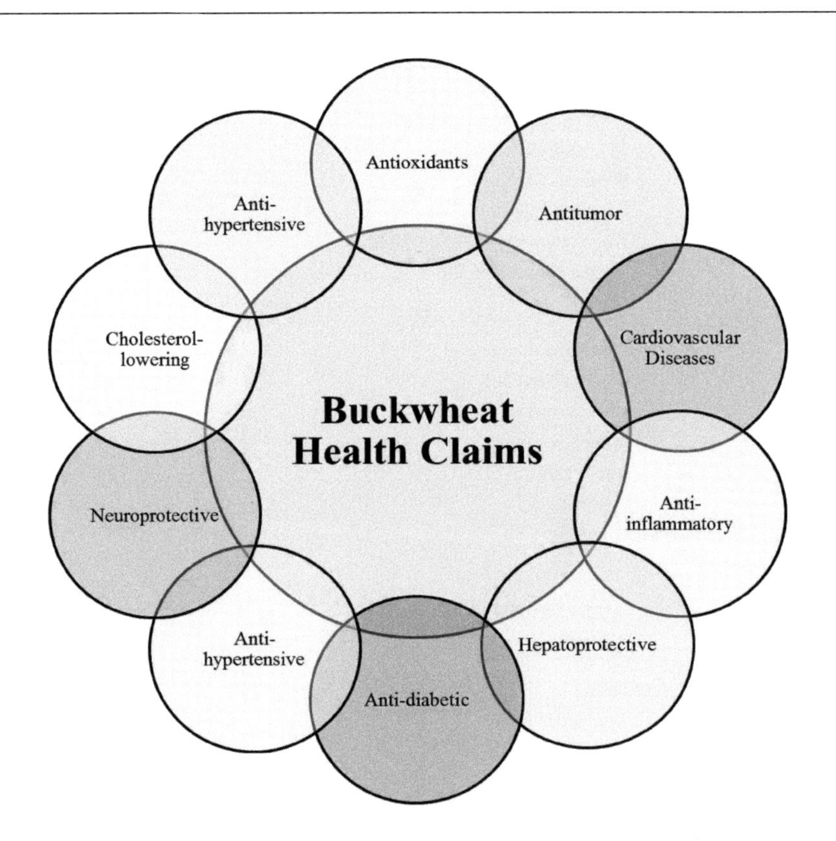

FIGURE 3.3 Various health claims for buckwheat.

TABLE 3.4
Buckwheat Metabolites and Their Medicinal Properties

Metabolites	Properties	References
Polysaccharides, procyanidin dimer, quercetin and tannins	Anti-tumor	Ji et al. (2019), Jing et al. (2016), Joshi et al. (2020), Lv et al. (2017), Wang et al. (2016)
Catechins, coumarins, rutin, curcuminoids, mandelic acid, lignans, polysaccharides, phenolic acids, quercetin, stilbenes and tannins	Anti-oxidant	Bose et al. (2018), Ji et al. (2019), Jing et al. (2016), Joshi et al. (2020), Lv et al. (2017), Tungmunnithum et al. (2018), Wang et al. (2018), Wang et al. (2016)
Apigenin, chrysin, hispidulin, hesperidin, isoorientin, isovitexin, luteolin, polysaccharides, quercetin and rutin	Anti-inflammatory	Bose et al. (2018), Giménez-Bastida and Zieliński (2015), Ji et al. (2019), Tungmunnithum et al. (2018), Wang et al. (2018)
Apigenin, naringenin, polysaccharides, quercetin, quercitrin, rutin and silymarin	Hepatoprotective	Bose et al. (2018), Ji et al. (2019), Jing et al. (2016), Joshi et al. (2020), Zhang et al. (2015)
Chlorogenic acid, epicatechin, hydroxybenzoic acid,luteolin, kaempferol, quercetin, quercitrin and rutin,	Anti-bacterial	Tungmunnithum et al. (2018), Wang et al. (2018)
Quercetin and esperidin	Anti-fungal	Bose et al. (2018)
Apigenin, catechin, emodin, rutin, dihydroquercetin, morin, quercetin and hesperidine	Anti-viral	Bose et al. (2018), Wang et al. (2018), Zhou et al. (2020)
Kaempferol, quercetin and rutin	Anti-ulcer	Bose et al. (2018)
Globulin	Anti-fatigue	Jing et al. (2016)
Polysaccharides	Hypolipidemic	Ji et al. (2019), Wang et al. (2016)
Apigenin, hesperidin, luteolin, polysaccharides and quercetin	Immunoregulatory	Ji et al. (2019), Tungmunnithum et al. (2018), Wang et al. (2016)
Galangin, kaempferol, myricetin, N-*trans*feruloyltyramine, polysaccharides and rutin	Neuroprotective	Giménez-Bastida andZieliński (2015), Ji et al. (2019), Jing et al. (2016), Wang et al. (2016), Wang et al. (2018), Zhu (2016)
D-*chiro*-inositol, isoquercetin, polysaccharides, quercetin and rutin	Anti-diabetic	Bose et al. (2018), Ji et al. (2019), Joshi et al. (2020), Lv et al. (2017), Wang et al. (2018), Zhang et al. (2015)
Apigenin, cinnamic acid, ferulic acid, gallic acid, isorhamnetin, kaempferol, luteolin, naringenin, quercetin, resveratrol, rutin and syringic acid	Cardioprotective	Bose et al. (2018), Gorniak et al. (2019), Joshi et al. (2020), Tungmunnithum et al. (2018), Wang et al. (2018)

(*Continued*)

TABLE 3.4 (CONTINUED)
Buckwheat Metabolites and Their Medicinal Properties

Metabolites	Properties	References
Apigenin, naringenin, nobiletin, Phenylpropanoid glycosides, procyanidin dimer, quercetin and tangeretin	Anti-cancer	Giménez-Bastida and Zieliński (2015), Gorniak et al. (2019), Jing et al. (2016), Lv et al. (2017), Wang et al. (2018), Tungmunnithum et al. (2018)
Quercetin	Anti-atherosclerosis	Bose et al. (2018), Gorniak et al. (2019)
Fagopyritol A1 and rutin	Blood glucose and cholesterol lowering	Giménez-Bastida and Zieliński (2015), Joshi et al. (2020), Wu et al. (2018), Lv et al. (2017)
Catechins, fisetin, genistein, hydrobenzoic acids, kaempferol, synigric acid, toxifolin and vanillic acid	Anti-neoplastic	Bose et al. (2018), Joshi et al. (2020)
Catechins, galangin, kaempferol and myricetin	Anti-aging	Joshi et al. (2020), Wang et al. (2018)
Kaempferol, myricetin, rutin and quercetin	Anti-thrombotic	Bose et al. (2018), Lv et al. (2017)

(Alvarez-Jubete et al. 2010). Buckwheat lignans have also shown promising results in reducing the size (50%) and number (37%) of mammary tumors in carcinogen-treated rats (Rickard and Thompson 2000). Kim et al. (2007) reported the anti-tumor activity of buckwheat is the result of its lowering serum estradiol, thus inhibiting mammary gland cancer. In another study, it was found that the consumption of buckwheat protein acts as a shielding agent to prevent colorectal cancer by lowering cell proliferation (Liu et al. 2008). Gao et al. (2007) isolated the novel anti-tumor protein (TBWSP31) from *F. tataricum* that has the capability to prevent proliferation of human mammary cancer Bcap37 cell line in a dose- and time-dependent way. Gulpinar et al. (2012) reported that ethanol and ethyl acetate extracts of buckwheat are known to possess neuroprotective effects due to acetyl cholinesterase, butyrylcholinesterase and tyrosinase inhibitory and antioxidant activity. The fagopyrins present in buckwheat play an important role in treating diabetes and cancer through photodynamic therapy as reported by various authors (Dai et al. 2009; Amezqueta et al. 2012; Tavcar et al. 2014).

CONCLUSION AND FUTURE POTENTIAL

The popularity of buckwheat has gained rapid momentum across the globe due to its high nutritional profile, gluten-free nature and other nutraceutical properties.

Therefore, current research should be focused on developing novel byproducts of buckwheat of commercial importance to cater to the demands of the population, especially in the higher Himalayan regions. Moreover, a multi-interdisciplinary approach is needed to improve buckwheat production and processing technology, enhance its bioactive constituents by unraveling various metabolomic pathways and testing its products on the market.

REFERENCES

Acanski, M., Pastor, K., Psodorov, D., Vujic, D., Razmovski, R. and Kravic, S. 2015. Determination of the presence of buckwheat flour in bread by the analysis of minor fatty acid methyl esters. *Adv Technol* 4(2): 86–92.

Ahmad, M., Ahmad, F., Dar, E. A., Bhat, R. A., Mushtaq, T. and Shah, F. 2018. Buckwheat (*Fagopyrum esculentum*) – A neglected crop of high altitude cold arid regions of Ladakh: Biology and nutritive value. *Intl J Pure Applbiosci* 6(1): 395–406.

Ahmed, A., Khalid, N., Ahmad, A., Abbasi, N. A., Latif, M. S. Z. and Randhawa, M. A. 2013. Phytochemicals and biofunctional properties of buckwheat: A review. *J Agric Sci* 152(3): 349–369.

Alencar, N. M. M. and Oliveria, L. C. 2019. Advances in pseudocereals: Crop cultivation, food application, and consumer perception. *Bioact Mol Food* 1695–1713.

Allen, S. and de Brauw, A. 2018. Nutrition sensitive value chains: Theory, progress, and open questions. *Glob Food Secur* 16: 22–28. doi:10.1016/j.gfs.2017.07.

Almedia, A. S. 1978. Cultura do trigosarraceno, Revista Paranaense de Desenvolvimento. *Curitiba* 62: 69–86.

Alvarez-Jubete, L., Arendt, E. K. and Gallagher, E. 2010. Nutritive value of pseudocereals and their increasing use as functional gluten-free ingredients. *Trends Food Sci Technol* 21(2): 106–113.

Amezqueta, S., Galán, E., Fuguet, E., Carrascal, M., Abián, J. and Torres, J. L. 2012. Determination of D-fagomine in buckwheat and mulberry by cation exchange HPLC/ ESI-Q-MS. *Anal BioanalyChem* 402(5): 1953–1960.

Awatsuhara, R., Harada, K., Maeda, T., Nomura, T. and Nagao, K. 2010. Antioxidative activity of the buckwheat polyphenol rutin in combination with ovalbumin. *Mol Med Rep* 3(1): 121–125.

Babu, S., Yadav, G. S., Singh, R., Avasthe, R. K., Das, A., Mohapatre, K. P., Tahashildar, M. et al. 2018. Production technology and multiferous uses of buckwheat (*Fagopyrum* spp.): A review. *Ind J Agron* 63(4): 415–427.

Barrett, C. B. 2010. Measuring food insecurity. *Science* 327(5967): 825–828. doi:10.1126/science.1182768.

Bastida, J. A. G. and Zielinski, H. 2015. Buckwheat as a functional food and its effects on health. *J Agric Food Chem* 63(36): 7896–7913.

Baumgertel, A., Grimm, R., Eisenbeiss, W. and Kreis, W. 2003. Purification and characterization of a flavonol 3-O-betaheterodisaccharidase from the dried herb of *Fagopyrum esculentum* Moench. *PhytoChem* 64: 411–418.

Becker, R. 2008. Fatty acids in food cereal grains and grain products. In: Chow, C. K. (ed.) *Fatty Acids in Foods and Their Health Implications*. CRC Press, Boca Raton FL, pp. 303–316.

Bharali, S. and Chrungoo, N. K. 2003. Amino acid sequence of the 26 kDa subunit of legumin-type seed storage protein of common buckwheat (*Fagopyrum esculentum* Moench): Molecular characterization and phylogenetic analysis. *Phytochem* 63: 1–5.

Bonafaccia, G., Marocchini, M. and Kreft, I. 2003. Composition and technological properties of the flour and bran from common and Tartary buckwheat. *Food Chem* 80(1): 9–15.

Borovaya, S. A. and Klykov, A. G. 2020. Some aspects of flavonoid biosynthesis and accumulation in buckwheat plants: Review. *Plant Biotechnol Rep* 14(2): 213–225. doi:10.1007/s11816- 676 020-00614-9.

Bose, S., Sarkar, D., Bose, A. and Mandal, S. S. 2018. Natural flavonoids and its pharmaceutical importance. *Pharm Rev*: 61–75.

Cai, E. P. and Lin, J. M. 2009. Epigallocatechingallate (EGCG) and rutin suppress the glucotoxicity through activating IRS2 and AMPK signaling in rat pancreatic β-Cells. *J Agric Food Chem* 57(20): 9817–9827.

Campbell, C. G. 1997. *Buckwheat Fagopyrum esculentum Moench. Promoting the Conservation and Use of Underutilized and Neglected Crops. 19.* IPK, Germany and IPGRI, Rome, Italy.

Choi, Y. S., Lee, H. H. and Park, C. H. 2003. Food, chemical and nutraceutical research on buckwheat in Korea: Literature survey. *Fagopyrum* 20: 73–80.

Christa, K. and Soral-Smietana, M. 2008. Buckwheat grains and buckwheat products – Nutritional and prophylactic value of their components – A review. *Czech J Food Sci* 26(3): 153–162.

Clemetson, C. A. 1976. Ascorbic acid and diabetes mellitus. *Med Hypo* 2(5): 193–194.

Comino, I., Moreno, M. L., Real, A., Rodriguez-Herrera, A., Barro, F. and Sousa, C. 2013. The gluten-free diet: Testing alternative cereals tolerated by celiac patients. *Nutrients* 5(10): 4250–4268.

Dai, T., Huang, Y. Y. and Hamblin, M. R. 2009. Photodynamic therapy for localized infections – State of the art. *Photodiagnphotodynther* 6(3–4): 170–188.

Dziedzic, K., Górecka, D., Szwengiel, A., Sulewska, H., Kreft, I., Gujska, E. and Walkowiak, J. 2018. The content of dietary fibre and polyphenols in morphological parts of buckwheat (*Fagopyrum tataricum*). *Plant Foods Hum Nutr* 73(1): 82–88. doi:10.1007/s11130-018-0659-0.

EASAC 2011. Plant genetic resources for foods and agriculture: Roles and research priorities in the European Union. *Policy Report 17-2011.*

FAO, IFAD and UNICEF and WHO 2017. *The State of Food Security and Nutrition in the World: Building Resilience for Peace and Food Insecurity.* FAO, Rome.

Fornal, Ł. 1999. Chemizmnasiongryki i kierunkispożywczegowykorzystania. *Biuletyn Naukowy* 4: 7–17.

Gabr, A. M. M., Sytar, O., Ghareeb, H. and Brestic, M. 2019. Accumulation of amino acids and flavonoids in hairy root cultures of common buckwheat (*Fagopyrum esculentum*). *Physiologyand Mol Biol Plants* 25(3): 787–797.

Gao, L., Li, X. Y., Zhang, Z., Wang, Z. H., Wang, H. W., Zhang, L. and Zhu, L. 2007. Apoptosis of HL-60 cells induced by recombinant common Buckwheat trypsin inhibitor. *J Exphematol* 15: 59–62.

Ge, R. H. and Wang, H. 2020. Nutrient components and bioactive compounds in Tartary buckwheat bran and flour as affected by thermal processing. *Intl J Food Prop* 23(1): 127–137. doi:10.1080/10942912.2020.1713151.

Giménez-Bastida, J. A. and Zieliński, H. 2015. Buckwheat as a functional food and its effects on health. *J Agric Food Chem* 63(36): 7896–7913. doi:10.1021/acs.jafc.5b02498.

Gimenez-Bastida, J. A., Piskuła, M. and Zielinski, H. 2015. Recent advances in processing and development of buckwheat derived bakery and non-bakery products – A review. *Pol J Food Nutrsci* 65(1): 9–20.

Gong, G., Qin, Y., Huang, W., Zhou, S., Yang, X. and Li, D. 2010. Rutin inhibits hydrogen peroxide–induced apoptosis through regulating reactive oxygen species mediated mitochondrial dysfunction pathway in human umbilical vein endothelial cells. *Eur J Pharmacol* 628(1–3): 27–35.

Gorniak, I., Bartoszewski, R. and Kroliczewski, J. 2019. Comprehensive review of antimicrobial activities of plant flavonoids. *Phytochem Rev* 18(1): 241–272.

Gulpinar, A. R., Orhan, I. E., Kan, A., Senol, F. S., Celik, S. A. and Kartal, M. 2012. Estimation of in vitro neuro-protective properties and quantification of rutin and fatty acids in buckwheat (*Fagopyrum esculentum*Moench) cultivated in Turkey. *Food Res Int* 46(2): 536–543.

Guo, X., Yao, H. and Chen, Z. 2007. Effect of heat, rutin and disulfide bond reduction on in vitro pepsin digestibility of Chinese Tartary buckwheat protein fractions. *Food Chem* 102(1): 118–122.

Han, S., Chu, J., Li, G., Zhu, L. and Shi, R. 2010. Effects of rutin from leaves and flowers of buckwheat (*Fagopyrum esculentum* Moench.) on angiotensin II–induced hypertrophy of cardiac myocytes and proliferation of fibroblasts. *Lat Am J Pharm* 29(1): 137–140.

Horbowicz, M. and Obendorf, R. L. 1992. Changes in sterols and fatty acids of buckwheat endosperm and embryo during seed development. *J Agric Food Chem* 40(5): 745–750.

Horbowicz, M., Brenac, P. and Obendorf, R. L. 1998. Fagopyritol B1, 0-α-d-galactopyranosyl-(1→2)-d-chiro-inositol, a galactosylcyclitol in maturing buckwheat seeds associated with desiccation tolerance. *Planta* 205(1): 1–11.

Ihme, N., Kiesewetter, H., Jung, F., Hoffmann, K. H., Birk, A., Muller, A. and Grutznar, K. I. 1996. Leg oedema protection from a buckwheat herbal tea in patients with chronic venous insufficiency: A single centre, randomized, double-blind, placebo-controlled clinical trial. *Eur J Clinpharmacol* 50: 443–447.

Ikeda, K. and Asami, Y. 2000. Mechanical characteristics of buckwheat noodles. *Fagopyrum* 17: 67–72.

Ikeda, S., Yamashita, Y., Tomura, K. and Kreft, I. 1999. Mineral composition of buckwheat by-products and its processing characteristics in konjak preparation. *Fagopyrum* 16: 89–94.

Ikeda, S., Yamashita, Y., Tomura, K. and Kreft, I. 2006. Nutritional comparison in mineral characteristics between buckwheat and cereals. *Fagopyrum* 23: 61–65.

Izydorczyk, M., Symons, S. J. and Dexter, G. E. 2002. Fractionation of wheat and barley. In: Marquart, L., Slavin, J. L. and Fulcher, R. (eds.) *Whole Grain Foods in Health and Disease*. St. Paul, MN: American Association of Cereal Chemists, pp. 47–82.

Janssen, F., Pauly, A., Rombouts, I., Jansens, K. J. A., Deleu, L. J. and Delcour, J. A. 2016. Proteins of amaranth (*Amaranthus* spp.), buckwheat (*Fagopyrum* spp.), and quinoa (*Chenopodium* spp.): A food science and technology perspective. *Compr Rev Food Sci Food Saf* 16(1): 39–58.

Ji, X., Han, L., Liu, F., Yin, S., Peng, Q. and Wang, M. 2019. A mini-review of isolation, chemical properties and bioactivities of polysaccharides from buckwheat (*Fagopyrum* Mill). *Int J Biol Macromol* 127: 204–209. doi:10.1016/j.ijbiomac.2019.01.043.

Jiang, P., Burczynski, F., Campbell, C., Pierce, G., Austria, J. A. and Briggs, C. J. 2007. Rutin and flavonoid contents in three buckwheat species *Fagopyrum esculentum*, *F. tataricum*, and *F. homotropicum* and their protective effects against lipid peroxidation. *Food Res Int* 40(3): 356–364.

Jing, R., Li, H., Hu, C., Jiang, Y., Qin, L. and Zheng, C. 2016. Phytochemical and pharmacological profiles of three *Fagopyrum* species. *Intl J Mol Sci* 17(4): 589. doi:10.3390/ijms17040589.

Joshi, D. C., Chaudhari, G. V., Sood, S., Kant, L., Pattanayak, A., Zhang, K., Fan, Y., Janovská, D., Meglič, V. and Zhou, M. 2019. Revisiting the versatile buckwheat: Reinvigorating genetic gains through integrated breeding and genomics approach. *Planta*. doi:10.1007/s00425-018-03080-4.

Joshi, D. C., Zhang, K., Wang, C., Chandora, R., Khurshid, M., Li, J., He, M., Georgiev, M. I. and Zhou, M. 2020. Strategic enhancement of genetic gain for nutraceutical development in buckwheat: A genomics-driven perspective. *Biotechnol Adv* 39: 107479. doi:10.1016/j.biotechadv.2019.107479.

Kalinova, J. P., Vrchotova, N. and Triska, J. 2019. Phenolics levels in different parts of common buckwheat (*Fagopyrum esculentum*) achenes. *J Cereal Sci* 85: 243–248. doi:10.1016/j.jcs.2018.12.012.

Kato, N., Kayashita, J. and Tomotake, H. 2001. Nutritional and physiological functions of buckwheat protein. *Recent Res Develop Nutr* 4: 113–119.

Kim, S. J., Maeda, T., Sarker, M. Z. I., Takigawa, S., Matsuura-Endo, C., Yamaguchi, H., Mukasa, Y., Saito, K., Hashimoto, N., Noda, T., Saito, T. and Suzuki, T. 2007. Identification of anthocyanins in the sprouts of buckwheat. *J Agric Food Chem* 55(15): 6314–6318.

Kim, S. L., Kim, S. K. and Park, C. H. 2002. Comparisons of lipid, fatty acids and tocopherols of different buckwheat species. *Food Sci Biotechnol* 11: 332–336.

Kim, S. L., Kim, S. K. and Park, C. H. 2004. Introduction and nutritional evaluation of buckwheat sprouts as a new vegetable. *Food Res Int* 37(4): 319–327.

Kitabayashi, H., Ujihara, A., Hirose, T. and Minami, M. 1995. On the genotypic differences for rutin content in Tatary buckwheat, *Fagopyrum tataricum* Gaertn. *Breed Sci* 45(2): 189–194.

Kreft, I., Zhou, M. L., Golob, A., Germ, M., Likar, M., Dziedzic, K. and Luthar, Z. 2020. Breeding buckwheat for nutritional quality. *Breed Sci* 70(1): 67–73. DOI:10.1270/jsbbs.19016.

Krkošková, B. and Mrazova, Z. 2005. Prophylactic components of buckwheat. *Food Res Int* 38(5): 561–568.

Kumar, A., Metwal, M., Kaur, S., Gupta, A. K., Puranik, S., Singh, S., Gupta, S. S., Babu, B. K., Sood, S. and Yadav, R. 2016. Nutraceutical value of finger millet [*Eleusinecoracana* (L.) Gaertn.], and their improvement using omics approaches. *Front Plant Sci* 7: 1–14.

Kuntic, V., Filipovic, I. and Vujic, Z. 2011. Effects of rutin and hesperidin and their Al (III) and Cu (II) complexes on in vitro plasma coagulation assays. *Molecules* 16(2): 1378–1388.

Kwon, S. J., Roy, S. K., Choi, J., Park, J., Cho, S., Sarker, K. et al. 2018. Recent research updates on functional components in buckwheat. *J Agric Sci-ChungbukNatlUni* 34(1): 1–8.

Landberg, R., Sun, Q., Rimm, E. B., Cassidy, A., Scalbert, A., Mantzoros, C. S., Hu, F. B. and Van-Dam, R. M. 2011. Selected dietary flavonoids are associated with markers of inflammation and endothelial dysfunction in U. S. women. *J Nutr* 141(4): 618–625.

Li, H. 2019. Buckwheat. In: Wang, J., Sun, B. and Tsao, R. (eds.), *Bioactive Factors and Processing Technology for Cereal Foods*. Springer Nature, Singapore, pp. 137–150.

Li, J., Yang, P., Yang, O., Gong, X., Ma, H., Dang, K., Chen, G., Gao, X. and Feng, B. 2019. Analysis of flavonoid metabolites in buckwheat leaves using UPLC-ESI-MS/MS. *Molecules* 24(7): 1310. doi:10.3390/molecules24071310.

Li, S. and Zhang, Q. H. 2001. Advances in the development of functional foods from buckwheat. *Crit Rev Food Sci Nutr* 41(6): 451–464.

Li, Z. X., Yu, X. Z., Zhang, L., Du, S. K. and Li, Y. P. 2001. Study of buckwheat flour usage in bread. *J Nutr* 2: 25–28.

Liu, C. L., Chen, Y. S., Yang, J. H. and Chiang, B. H. 2008. Antioxidant activity of Tartary (*Fagopyrum tataricum* (L.) Gaertn.)and Common (*Fagopyrum esculentum* Moench) Buckwheat Sprouts. *J Agric Food Chem* 56(1): 173–178.

Lv, L., Xia, Y., Zou, D., Han, H., Wang, Y., Fang, H. and Li, M. 2017. *Fagopyrum tataricum* (L.)Gaertn.: A review on its traditional uses, phytochemical and pharmacology. *Food Sci Technol Res* 23(1): 1–7.

Ma, J. F., Zheng, S. J., Matsumoto, H. and Hiradate, S. 1997. Detoxifying aluminium with buckwheat. *Nature* 390(6660): 569–570.

Mabhaudhi, T., Chimonya, V. G. P., Hlahla, S., Massawe, F., Mayes, S., Modi, A. T. and Nahmo, L. 2019. Prospects of orphan crops in climate change. *Plantahttps.* doi:10.1007/s00425-019-03129-y.

Martin-Garcia, B., Pasini, F., Verardo, V., Gomez-Caravaca, A. M., Marconi, E. and Caboni, M. F. 2019. Distribution of free and bound phenolic compounds in buckwheat milling fractions. *Foods* 8(12): 670. DOI:10.3390/foods8120670.

Massawe, F., Mayes, S., Cheng, A., Chai, H., Cleasby, P., Symonds, R. C., Ho, W., Siise, A., Wong, Q., Kendabie, P., Yanusa, Y., Jamalluddin, N., Singh, A., Azman, R. and Azam-Ali, S. N. 2015. The potential for underutilised crops to improve food security in the face of climate change. *Proc Environ Sci* 29: 140–141.

Matsui, K. and Walker, A. R. 2019. Biosynthesis and regulation of flavonoids in buckwheat. *Breed Sci* 1–11. DOI:10.1270/jsbbs.19041.

Melini, V., Melini, F. and Acquistucci, R. 2020. Phenolic compounds and bioaccessibility thereof in functional pasta: Review. *Antioxidants* 9(343). doi:10.3390/antiox9040343.

Milisavljevic, M. D., Timotijevic, G. S., Radovic, S. R., Brkljacic, J. M., Konstantinovic, M. M. and Maksimovic, V. R. 2004. Vicilin-like storage globulin from buckwheat (*Fagopyrum esculentum*Moench) seeds. *J Agric Food Chem* 52(16): 5258–5262.

Mohajan, S., Munna, M. M., Orchy, T. N., Hoque, M. M. and Farzana, T. 2019. Buckwheat flour fortified bread. *Bangla J Sci Indus Res* 54(4): 347–356.

Nam, T. G., Lim, Y. J. and Eom, S. H. 2018. Flavonoid accumulation in common buckwheat (*Fagopyrum esculentum*) sprout tissues in response to light. *Hortic Environbiotechnol* 59(1): 19–27. doi:10.1007/s13580-018-0003-5.

Ohsawa, R. and Tsutsumi, T. 1995. Inter-varietal variations of rutin content in common buckwheat flour (*Fagopyrum esculentum* Moench). *Euphyticam* 86: 183–189.

Park, B. I., Kim, J., Lee, K., Lim, T. and Hwang, K. T. 2019. Flavonoids in common and Tartary buckwheat hull extracts and antioxidant activity of the extracts against lipids in mayonnaise. *J Food Sci Technol* 56(5): 2712–2720. doi:10.1007/s13197-019-03761-2.

Park, C. H., Yeo, H. J., Park, Y. J., Morgan, A. M. A., Arasu, M. V., Al-Dhabi, N. A. and Park, S. U. 2017. Influence of indole-3-acetic acid and gibberellic acid on phenylpropanoid accumulation in common buckwheat (*Fagopyrum esculentum* Moench) sprouts. *Molecules* 22(3). doi:10.3390/molecules22030374.

Park, N. I., Li, X., Uddin, M. R. and Park, S. U. 2011. Phenolic compound production by different morphological phenotypes in hairy root cultures of *Fagopyrum tataricum* Gaertn. *Arch Biol Sci* 63(1): 193–198.

Pirzadah, T. B., Malik, B., Tahir, I. and Rehman, R. U. 2013. Buckwheat: An introspective and future perspective with reference to Kashmir Himalayas. In *Proceedings of 12th International Symposium Laško, Slovania. Proceedings of International Symposium*, August 21–25, pp. 212–215.

Pirzadah, T. B., Malik, B., Tahir, I., Qureshi, M. I. and Rehman, R. U. 2017. Metabolite fingerprinting and antioxidant potential of Tartary buckwheat – an underutilized pseudocereal crop from Kashmir region. *Free Rad Antioxid* 7(1): 95–106.

Pirzadah, T. B., Malik, B., Tahir, I. and Rehman, R. U. 2018. Antioxidant potential and ionomic analysis of two buckwheat species from Kashmir region. *Pharmacog J* 10(6): s83–s88.

Pirzadah, T. B., Malik, B., Tahir, I. and Rehman, R. U. 2020. Buckwheat journey to functional food sector. *Currnutr Food Sci* 16(2): 134–141.

Podolska, G. 2016. The effect of habitat conditions and agrotechnical factors on the nutritional value of buckwheat. In: Zhou, M., Kreft, I., Woo, S. H., Chrungoo, N. and Wieslander, G. (eds.) *Molecular Breeding and Nutritional Aspects of Buckwheat.* Academic Press, Cambridge, pp. 283–297.

Qian, J. and Kuhn, M. 1999. Physical properties of buckwheat starches from various origins. *Starch Starke* 51(2–3): 81–85.

Qin, P., Tingjun, M., Li, W., Fang, S. and Guixing, R. 2011. Identification of Tartary buckwheat tea aroma compounds with gas chromatography–mass spectrometry. *J Food Sci* 76: 401–407.

Radović, S. R., Maksimović, V. R. and Varkonji-Gasić, E. I. 1996. Characterization of buckwheat seed storage proteins. *J Agric Food Chem* 44(4): 972–974.

Radović, S. R., Maksimović, V. R. and Varkonji-Gasić, E. I. 1999. Characterization of buckwheat seed storage proteins. *J Agric Food Chem* 44(4): 972–974.

Ramos-Romero, S., Hereu, M., Atienza, L., Amézqueta, S., Casas, J., Muñoz, S., Medina, I., Miralles-Pérez, B., Romeu, M. and Torres, J. L. 2020. The buckwheat iminosugar d-Fagomine attenuates sucrose-induced steatosis and hypertension in rats. *Mol Nutr Food Res* 64(1). doi:10.1002/mnfr.201900564.

Rickard, S. E. and Thompson, L. U. 2000. Urinary composition and posprandial blood changes in H-secoisolariciresinoldiglycoside metabolites in rats do not differ between acute and chronic SDG treatments. *J Nutr* 130(9): 2299–2305.

Rozanska, D., Mikos, K. and Regulska-Ilow, B. 2020. Assessment of the glycemic index of groats available on the polish food market. *Rocz Panstw Zakl Hig* 71(1): 81–87. doi:10.32394/rpzh.2020.0101.

Ruan, J., Zhou, Y., Yan, J., Zhou, M., Woo, S. H., Weng, W., Cheng, J. and Zhang, K. 2020. Tartary buckwheat: An under-utilized edible and medicinal herb for food and nutritional security. *Food Rev Intl.* doi:10.1080/87559129.2020.1734610.

Sharmila, G., Bhat, F. A., Arunkumar, R., Elumalai, P., Singh, P. R., Kumar, K. S. and Arunakaran, J. 2014. Chemopreventive effect of quercetin, a natural dietary flavonoid on prostate cancer in *in vivo* model. *Clinnutr* 33(4): 718–726.

Sinkovic, L., Kokalj, D., Vidrih, R. and Meglic, V. 2020. Milling fractions fatty acid composition of common (*Fagopyrum esculentum* Moench) and Tartary (*Fagopyrum tataricum* (L.) Gaertn) buckwheat. *J Stored Prod Res* 85. doi:10.1016/j.jspr.2019.101551. http://www.ncbi.nlm.nih.gov/pubmed/101551.

Skerritt, J. H. 1986. Molecular comparison of alcohol-soluble wheat and buckwheat proteins. *Cereal Chem* 63: 365–369.

Skrabanja, V. and Kreft, I. 2016. Nutritional value of buckwheat proteins and starch. In: Zhou, M. L., Kreft, I., Woo, S. H., Chrungoo, H. and Wieslander, G. (eds.) *Molecular Breeding and Nutritional Aspects of Buckwheat.* Academic Press is an imprint of Elsevier, UK, pp. 169–176.

Stibilj, V., Kreft, I., Smrkolj, P. and Osvald, J. 2004. Enhanced selenium content in buckwheat (*Fagopyrum esculentum* Moench) and pumpkin (*Cucurbitapepo* L.) seeds by foliar fertilization. *Eur Food Res Technol* 219: 142–144.

Subedi, N. 2018. *Changes in Phytochemical Properties of Buckwheat Varieties on Malting* (Thesis). Department of Food Technology, Tribhuvan University, Nepal.

Suzuki, T., Noda, T., Morishita, T., Ishiguro, K., Otsuka, S. and Brunori, A. 2020. Present status and future perspectives of breeding for buckwheat quality. *Breed Sci* 1–19. doi:10.1270/jsbbs.19018.

Sytar, O., Biel, W., Smetanska, I. and Brestic, M. 2018. Bioactive compounds and their biofunctional properties of different buckwheat germplasms for food processing. In: Zhou, M., Kreft, I., Tang, Y. and Suvorova, G. (eds.) *Buckwheat Germplasm in the World*, 1st edn. Elsevier Publications, USA, pp. 191–204.

Taguchi, G. 2016. Flavonoid biosynthesis in buckwheat. In: Zhou, M. L., Kreft, I., Woo, S. H., Chrungoo, N. and Wieslander, G. (eds.) *Molecular Breeding and Nutritional Aspects of Buckwheat*. Academic Press is an imprint of Elsevier, UK, pp. 377–386.

Tavcar Benkovic, E., Zigon, D., Friedrich, M., Plavec, J. and Kreft, S. 2014. Isolation, analysis and structures of phototoxic fagopyrins from buckwheat. *Food Chem* 143: 432–439.

Thwe, A. A., Kim, J. K., Li, X., Kim, Y. B., Uddin, M. R., Kim, S. J., Suzuki, T., Park, N. I. and Park, S. U. 2013. Metabolomic analysis and phenylpropanoid biosynthesis in hairy root culture of Tartary buckwheat cultivars. *PLOS ONE* 8(6): e65349.

Tien, N. N. T., Trinh, L. N. G., Inoue, N., Morita, N. and Hung, P. V. 2018. Nutritional composition, bioactive compounds, and diabetic enzyme inhibition capacity of three varieties of buckwheat in Japan. *Cereal Chem* 95(5): 615–624. doi:10.1002/cche.10069.

Tolaini, V., Fiore, A. D., Nobili, C., Presenti, O., De Rossi, P., Procacci, S., Vitali, F. and Brunori, A. 2016. Exploitation of Tartary buckwheat as sustainable ingredient for healthy foods production. *AgricAgric Sci Procedia* 8: 455–460.

Tuan, P. A., Thwe, A. A., Kim, J. K., Kim, Y. B., Lee, S. and Park, S. U. 2013. Molecular characterisation and the light–dark regulation of carotenoid biosynthesis in sprouts of Tartary buckwheat (*Fagopyrum tataricum* Gaertn.). *Food Chem* 141(4): 3803–3812.

Tungmunnithum, D., Thongboonyou, A., Pholboon, A. and Yangsabai, A. 2018. Flavonoids and other phenolic compounds from medicinal plants for pharmaceutical and medical aspects: An overview. *Medicines* 5(3): 93. doi:10.3390/medicines5030093.

Wang, Q., Takao Ogura, T. and Wang, L. 1995. Research and development of new products from Bitter-Buckwheat. *Curr Adv Buckwheat Res* 873–879.

Wang, T., Li, Q. and Bi, K. 2018. Bioactive flavonoids in medicinal plants: Structure, activity and biological fate: Review. *Asian J Pharm Sci* 13(1): 12–23.

Wang, X. T., Zhu, Z. Y., Zhao, L., Sun, H. Q., Meng, M., Zhang, J. Y. and Zhang, Y. M. 2016. Structural characterization and inhibition on alpha-D-glucosidase activity of non-starch polysaccharides from *Fagopyrum tartaricum*. *Carbohydrpolym* 153: 679–685.

Wang, Y. and Campbell, C. G. 2004. Buckwheat production, utilization, and research in China. *Fagopyrum* 21: 123–133.

Wijngaard, H. H. and Arendt, E. K. 2006. Buckwheat. *Cereal Chem* 83(4): 391–401.

Wu, W., Wang, L., Qiu, J. and Li, Z. 2018. The analysis of fagopyritols from Tartary buckwheat and their anti-diabetic effects in KK-Ay type 2 diabetic mice and HepG2 cells. *Jfunct Foods* 50: 137–146.

Yilmaz, M., Kantarjian, H., Wang, X., Khoury, J. D., Ravandi, F., Jorgensen, J. et al. 2020. The early achievement of measurable residual disease negativity in the treatment of adults with Philadelphia-negative B-cell acute lymphoblastic leukemia is a strong predictor for survival. *Am J Hematol* 95(2): 144–150. doi:10.1002/ajh.25671.

Zhang, X., Huang, H., Zhao, X., Lv, Q., Sun, C., Li, X. and Chen, K. 2015. Effects of flavonoids-rich Chinese bayberry (MyricarubraSieb. etZucc.) pulp extracts on glucose consumption in human HepG2 cells. *J Funct Foods* 14: 144–153.

Zhang, Z. L., Zhou, M. L., Tang, Y., Li, F. L., Tang, Y. X., Shao, J. R., Xue, W. T. and Wu, Y. M. 2012. Bioactive compounds in functional buckwheat food. *Food Res Int* 49(1): 389–395.

Zhou, M. L., Kreft, I., Woo, S. H., Chrungoo, N. and Wieslander, G. 2016. Bioactive compounds in buckwheat sprouts. In: Zhou, M. L., Kreft, I., Woo, S. H., Chrungoo, N. and Wieslander, G. (eds.) *Molecular Breeding and Nutritional Aspects of Buckwheat.* Academic Press is an imprint of Elsevier, UK, pp. 151–159.

Zhou, X., Hao, T., Zhou, Y., Tang, W., Xiao, Y., Meng, X. and Fang, X. 2015. Relationships between antioxidant compounds and antioxidant activities of Tartary buckwheat during germination. *J Food Sci Tech* 52(4): 2458–2463.

Zhou, Y., Hou, Y., Shen, J., Huang, Y., Martin, W. and Cheng, F. 2020. Network-based drug repurposing for novel coronavirus 2019-nCoV/SARS-CoV-2. *Cell Discov* 6: 14. doi:10.1038/s41421-020-0153-3.

Zhu, F. 2016. Chemical composition and health effects of Tartary buckwheat. *Food Chem* 203: 231–245.

4 Buckwheat: A Future Crop to Combat Abiotic Stress

INTRODUCTION

ABIOTIC STRESS IN BUCKWHEAT – A PERSPECTIVE

The unprecedented increase in human population, industrialization and urbanization has led to a myriad of environmental problems with debilitating consequences for the life support systems of the earth (Zhu 2016; Waqas et al. 2017; Zafar et al. 2018; Hasegawa et al. 2018; Wang et al. 2019). Plants being sessile face multiple abiotic stresses (salinity, cold, heat, drought, water and heavy metal [HM] stress), which cause an approximately 70% decline in crop production and thus present a threat to future food security (Mantri et al. 2012; Vaughan et al. 2018; Lohani et al. 2020). One of the concerns of contemporary society is the environmental degradation of land and water bodies with a variety of metals (particularly HMs). HM toxicity has put further constraints on an already shrinking land area for the cultivation of underutilized crops including neglected pseudocereals (Pirzadah et al. 2020). The first and foremost response of plants towards HM toxicity is the generation of reactive oxygen species (ROS), viz. hydrogen peroxide (H_2O_2), superoxide radical ($O_2^{\bullet -}$), hydroxyl radical ($OH^{\bullet -}$) due to intervention in the electron transport chain, particularly those of chloroplastic membranes (La-Rocca et al. 2009). The increased production of ROS leads to oxidative damage in plants that causes membrane disruption, lipid peroxidation, electrolyte leakage, DNA damage, apoptosis and ultimately the death of the plant (Rascio and Navari-Izzo 2011; Pirzadah et al. 2015). Nature has given plants a potential defense mechanism to combat these oxidative stresses, and this is often considered as a "green liver" that acts as a natural filter for these toxic chemicals. Plants possess an innate capacity to resist oxidative stresses, and the basic mechanisms involve the coordination of complicated physiological and biochemical processes including alterations at the gene level (Dalcorso et al. 2010). Moreover, an in-depth understanding at the cellular and molecular level is required to unravel the various factors that alter the homeostatic equilibrium and identify novel genes/metabolites or networks that play a vital role in the stress signal pathways (Hoque et al. 2018; Nadarajah 2020). A number of plants have been studied to date to understand their low to moderate tolerance mechanisms in a variety of environments, such as fluctuations in photoperiod; light intensity and quality; starvation and nutrient abundance; flooding and drought; salinity; and alterations in air, temperature and soil pollution, including exposure to HMs and osmotic changes (Apel and Hirt 2004; Canter 2018; Zorb et al. 2019). Pseudocereals certainly have strong mechanisms and better gene regulation to combat stress. Buckwheat, an important pseudocereal crop, not only thrives well in harsh environmental conditions but is also considered to be an important functional food. To ensure constant monitoring of various environmental parameters and a rapid and appropriate response, these neglected crops must have developed elaborated robust mechanisms of signal perception and transduction followed by the development of excellent tolerance strategies. However, for a number of plants, the significance of the ability to adapt to fluctuating environmental conditions has been reviewed in a number of research/review articles (Hirayama and Shinozaki 2010). Pseudocereals are currently gaining rapid momentum as nutraceutical foods due to their rich nutritional profile, and thus mass production of these lost crops on marginal lands is essential (Kumari and Chaudhary 2020). For profitable cultivation of pseudocereals on these polluted marginal lands, it is imperative to undertake detailed studies to investigate the plant responses to various environmental stresses (particularly HMs). Considerable progress has been made in the field of plant stress physiology, particularly for the development of stress-tolerant crop varieties that can withstand harsh climatic conditions. Though in the past, high-yielding crop varieties were developed using traditional breeding approaches, the knowledge that has been generated is limited to crossing-over and genetic recombination. Recently, genetic engineering for developing stress-tolerant crops has been focused mainly on the gene expressions that are associated with osmolyte biosynthesis (proline, glycine betaine, sugar); genes that encode various defensive enzymes (superoxide dismutase, glutathione reductase, catalase, ascorbate reductase) for the scavenging of reactive oxygen species (ROS); genes encoding late embryogenesis abundant proteins (*LEA*), dehydrins etc.; genes encoding heterologous enzymes with distinct temperature optima; genes for molecular chaperons (heat shock proteins [Hsps]); transcription factor genes (*DREB1A*, *CBF 1*); cell membrane engineering; and proteins associated with ion homeostasis (Lee et al. 2007; Atkinson and Urwin 2012; Parmar et al. 2017; Lohani et al. 2020). These engineering approaches would indeed unravel the novel pathways to develop climate-resilient crops with improved agronomic traits. While a lot of information pertaining to plant stress physiology has been reviewed, the mechanisms involved in detecting stress have largely remained elusive. Surprisingly, there is a huge knowledge gap as to why the stress-tolerant capacities of neglected crops have been ignored at large for their molecular analysis and tolerance mechanisms. In this chapter, we will address how buckwheat could be a potential candidate to combat abiotic stress constraints by catching up and connecting with modern developments in buckwheat research.

HEAVY METAL DETOXIFICATION MECHANISM IN BUCKWHEAT

Buckwheat possesses an excellent capacity to accumulate HMs without exhibiting any lethal morphological symptoms. Thus, besides being used as a nutraceutical food crop, it could also serve as a prime candidate in phytoremediation technology as it has a great tendency to accumulate toxic metal ions like aluminum (Al), lead (Pb) and mercury (Hg) (Pirzadah et al. 2019, 2020; Hakeem et al. 2019; Pirzadah et al. 2018) Buckwheat grows even in acidic soils because of its tendency to tolerate Al-stress (Ma et al. 1997). However, to better understand the nature of these tolerance mechanisms, it is essential to focus on the research pertaining to plant stress biology. Current research has unraveled the role of certain organic acids to detoxify the Al by either forming chelating complexes inside cytosol or at the root-soil interface or rhizosphere. Several plants are reported that ooze out certain organic acids when subjected to Al stress (Ryan et al. 1993). These include oxalic acid, malic acid and citric acid anions, which form firm complexes with Al thus acting as a shielding agent to protect roots. Production of these organic acid anions in response to Al stress depends upon the plant species; for example, citric acid is released by maize (Pellet et al. 1995) and soybeans (Yang et al. 2001), while buckwheat secretes oxalic acid (Ma et al. 1997). However, a few species like radish and rye secrete both malic and citric acid (Li et al. 2002). Detoxification of Al occurs either inside the root cells or in the rhizosphere region. The release of these organic acids is mainly localized at the apex of the root, as the apex of the root is sensitive to Al stress (Ryan et al. 1993); thus only the cations which immediately surround the cells of the apical root must be detoxified. Thus, it is sensible to impede the discharge of the organic acids particularly in the apical zone as it reduces the metabolic cost of the Al-tolerance mechanism. Thus, secretion needs to continue as the root apex moves forward through the acidic soil to replace the organic acids that get dispersed away from the root or broken down by microorganisms. It means that organic acids act as shielding agents to protect the root apex from Al toxicity. In buckwheat, organic acid anions are predominant inside the leaves. The secretion of these acid anions from the roots represents the external detoxification of the Al metal ions, thus giving buckwheat its high Al tolerance. Two patterns of secretion of organic acids have been proposed in terms of the time required in plants. In pattern I, the release of organic acid anions is quick when the plant is subjected to Al stress without any discernable delay; that is, the secretion of organic acid from buckwheat is an instantaneous phenomenon (15–30 minutes) when the plant is subjected to Al stress. However, in pattern II, the secretion of organic acids is deferred for several hours when the plant is subjected to Al stress, thus indicating that gene activation is required (Ma and Hiradate 2000) (Figure 4.1). Moreover, in buckwheat, there occur certain shifts in chemical reactions during uptake, translocation and accumulation. When buckwheat is subjected to Al stress, Al gains entry into the root cells due to inwardly directed electrochemical gradients for this ion; however, the exact mechanism is not known as yet. Inside, the root cells, it gets chelated by oxalic acid anion resulting in the formation of 1:3 Al-oxalate chelating complex (Ma et al. 1997). Upon translocation from root to shoot there occur shifts in chemical reaction i.e, ligand exchange reaction in the xylem to form Al-citrate complex (Ma and Hiradate 2000). During the translocation from xylem to leaf cells, another shift occurs in chemical reaction to reform the Al-oxalate chelating complex, which is then stored inside the vacuole (Ma and Hiradate 2000) (Figure 4.2). Shen and Ma (2001)

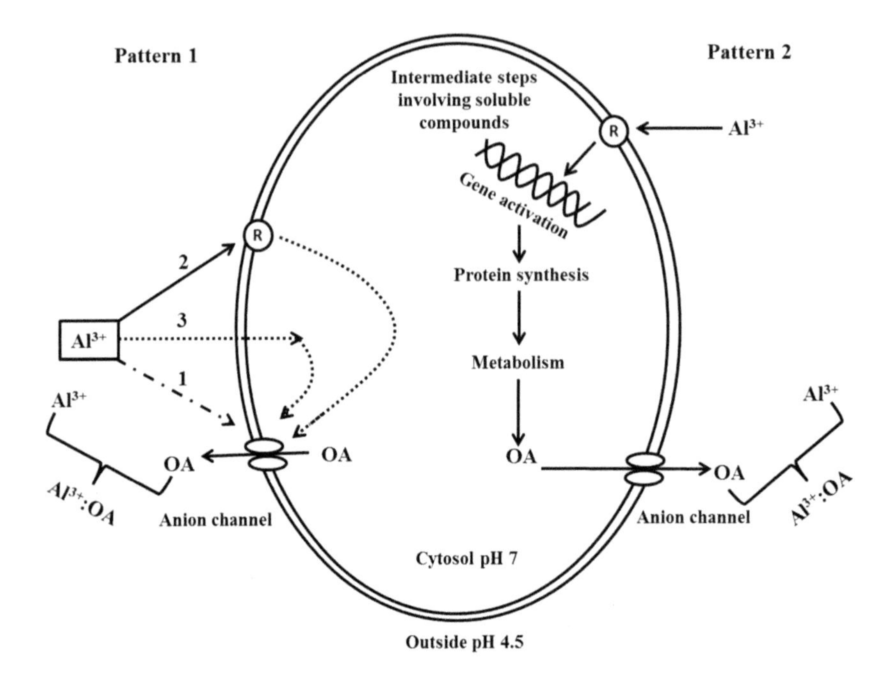

FIGURE 4.1 Models for the aluminum (Al)-stimulated secretion of organic acid anions (OA) from plant roots. For Pattern I-type responses, Al activates an anion channel on the plasma membrane that is permeable to organic acid anions. This stimulation could occur in one of three ways: (1) Al^{3+} interacts directly with the channel protein to trigger its opening; (2) Al^{3+} interacts with a specific receptor (R) on the membrane surface or with the membrane itself to initiate a secondary-messenger cascade that then activates the channel; or (3) Al^{3+} enters the cytoplasm and activates the channel directly or indirectly via secondary messengers. The Al-activated efflux from maize probably occurs by mechanism 1; the mechanism activating malate efflux from wheat is not known. In the Pattern II response, Al interacts with the cell, perhaps via a receptor protein (R) on the plasma membrane, to activate the transcription of genes that encode proteins involved with the metabolism of organic acids or their transport across the plasma membrane. Organic acid anions form a stable complex with Al, thereby detoxifying Al^{3+} in the rhizosphere. Experiments have identified some of the components shown in the model for Pattern I, whereas the components depicted for Pattern II are entirely speculative.

conducted an experiment to determine the mobility and distribution of Al in *F. esculentum* and reported that the mobility of the Al gets hindered after it gets accumulated in the leaves of buckwheat. Horbowicz et al. (2011) studied the impact of Al uptake and accumulation, as well as that of basic elements, on the content of anthocyanins in the seedlings of common buckwheat. Their research revealed that more Al concentration was accumulated in the cotyledons and hypocotyls of buckwheat seedlings than in whole buckwheat, confirming that buckwheat possesses an efficient Al-detoxification system. Moreover, they also concluded that high Al levels cause a decline in potassium accumulation in

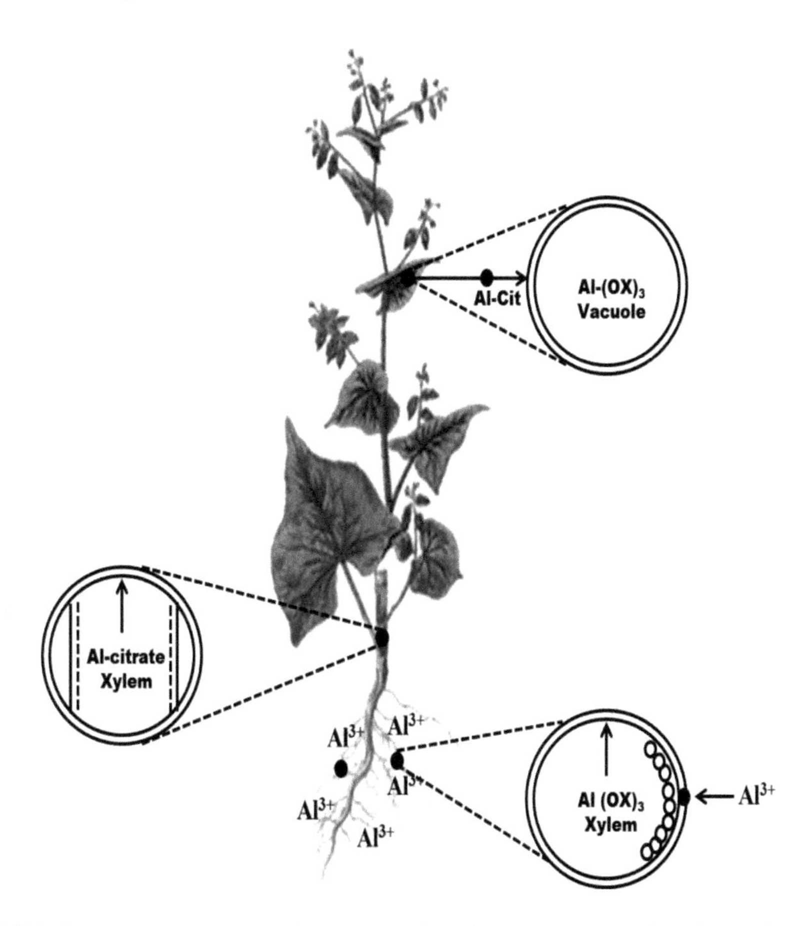

FIGURE 4.2 Uptake and distribution of various aluminum (Al) forms in the Al-accumulating buckwheat plant. Aluminum enters the root by an unknown mechanism, probably as Al^{3+}. Once it crosses the plasma membrane, the Al^{3+} is chelated with oxalate to form a 1:3 Al-oxalate complex. When Al is translocated from the roots to the shoots, a ligand-exchange reaction occurs in the xylem to form Al-citrate. Once unloaded from the xylem to leaf cells, another ligand exchange reaction occurs to reform the Al-oxalate complex, which is then stored in the vacuole. Abbreviations: Cit, citrate; Ox, oxalate; PM, plasma membrane (unpublished work of the author).

buckwheat hypocotyls and cotyledons, and a significant decline in calcium uptake by hypocotyls but not by cotyledons. Further, they also showed that higher levels of Al cause an increase in anthocyanin concentration in hypocotyl that confirms an additional mechanism of Al-detoxification based on forming complex compounds between anthocyanins and Al. Zhu et al. (2015) carried out the genome-wide transcriptomics and phylogenetic analysis and revealed that the basal root region and root tip of *F. tataricum* may have both common and distinct mechanisms of Al responsiveness and that organic acid metabolism is not a rate-limiting step for the secretion of organic acids induced by Al in buckwheat. Their results also suggested that xylem loading of Al may be the rate-limiting step for the translocation of Al from root to shoots in buckwheat and that two putative citrate transporters (*FtFRDL1* and *FtFRDL2*) may be prescribed for the translocation of Al through the release of citrate into the xylem for complexion with Al. They also proposed that the duplication and sub-functionalization of key genes in buckwheat coordinate with the Al accumulation and Al tolerance. In addition, in removing Al from the soil, buckwheat can be categorized as an efficient lead (Pb) hyperaccumulator (1000 mg kg^{-1} of Pb on dry-mass basis) (Pirzadah et al. 2015, 2020), particularly in the shoots. Previous studies also reported that *F. esculentum* has the ability to accumulate more Pb content, especially in its leaves (8000 mg kg^{-1} DW), followed by the stem (8000 mg kg^{-1} DW) and roots (3300 mg kg^{-1} DW), without exhibiting any symptoms, and these findings could be helpful for using the buckwheat to remediate Pb-contaminated soils (Tamura et al. 2005; Pirzadah et al. 2015, 2020). However, the Pb accumulation capacity in buckwheat can be improved (more than five times) with the addition of biodegradable chelator (20 mmol kg^{-1} methylglycinediacetic acid trisodium salt), as reported by Tamura et al. (2005). Horbowicz et al. (2013) studied the impact of short-term exposure to Pb and Cd on the flavonoid content and seedling growth of common buckwheat cultivar and suggested that low levels of Pb and Cd result in stimulation of seedling growth but that higher levels show the reverse effect. However, Cd stress was more severe than Pb stress, suggesting that the detoxification mechanism of Pb and Cd in buckwheat seedlings probably involves flavonoids. It has also been reported that buckwheat exhibited an efficient Pb and Hg detoxification mechanism as evidenced by its strong antioxidative defense system and enhanced production of osmoprotectants; it thus could be an ideal candidate for remediating soils contaminated with Pb, Al and Hg (Pirzadah et al. 2018, 2019, 2020; Wang et al. 2020). Wang et al. (2020) revealed that when buckwheat is subjected to Pb stress, the Pb ions translocate to the leaves and some of the Pb ions bind to the cell wall. Moreover, the majority of the ions translocate into the cytoplasmic matrix via various transporters, viz. cation diffusion facilitators (CDFs), Nramps and other ion transporters via iron or water channels (Hall and Williams 2003). After Pb ions gain entry to the cells, there is a fluctuation in the hormonal concentration that leads to the enhanced production of reduced glutathione. These hormones in turn interact with ROS and thus participate in the MAPK cascades that will switch on the transcription factors (TFs). Subsequently, the genes that exhibit modification in response to Pb stress are then regulated by TFs to defend against and alleviate Pb stress. Besides, the cells also maintain homeostasis through antioxidative enzyme defense system in all organelles. In this way, the Pb ions, after being chelated by GSH, phytochelatins (PHs) or their compounds, are transported into the vacuoles via MTPs, vacuolar cation/proton exchanger (VCE) (a homologue of YCF) and other related transporters. Thus, the findings of this study suggested the involvement of the metabolic processes of Tartary buckwheat in its response to Pb stress as well as showing how the transport and accumulation of Pb ions takes place in Tartary buckwheat leaves (Figure 4.3).

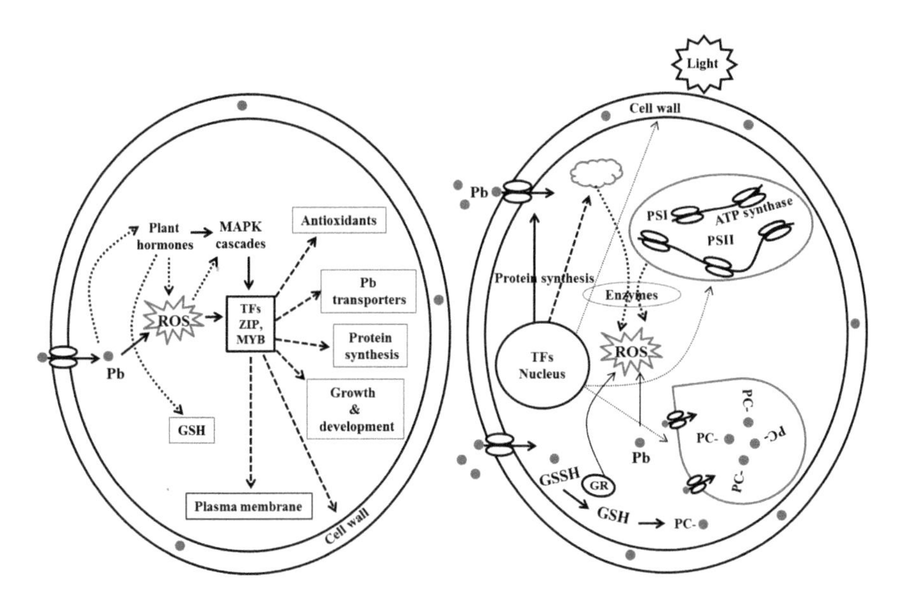

FIGURE 4.3 Schematic diagram of the response of Tartary buckwheat leaves to Pb-induced stress.

OMICS TOOLS – NEW INSIGHTS TO UNRAVEL BUCKWHEAT HEAVY METAL TOLERANCE

To address the challenges of HM toxicity and get a better understanding of plant–heavy metal dynamic interactions, the novel approaches of omics technologies play a significant role in unraveling the dynamic regulatory elements that help to impart stress tolerance in plants; moreover, the genes and metabolites could be used as novel candidates in crop improvement programs to engineer plants in enhancing tolerance to HM stress and developing crops with novel and desired agronomic traits. Current omics tools, which involve genomics, proteomics, ionomics, transcriptomics and metabolomics, help us to provide a comprehensive picture of plant–HM stress tolerance mechanisms. Over the past decade, extensive research on plant stress physiology has been carried out to unravel the signaling pathway mechanisms. While the genomics approach helps to address the abiotic stress responses in plants, the modifications in gene expression at transcript level have not always been reflected at protein level (Gygi et al. 1999; Bohnert et al. 2006). Proteomics plays a vital role in the identification and quantification of novel proteins/metabolites/pathways or genes that are involved in regulating the various abiotic stresses and provide us better insight into HM stress tolerance in plants, as well as an understanding the metabolic profiles of various plant parts employing metabolomics. Investigation of protein network regulation and interactions at the genetic or physiological level may be viewed via proteomic pathways and functional genomics. There is a long tradition of and thorough knowledge about proteome analysis, and indeed, proteome analysis can be better inferred by differentiating among various levels on the basis of its objectives. First, comparable to metabolite target analysis that utilizes specific methods for the analysis of analytes, viz. phytohormones, protein profiling would also provide relevant information. Second, protein profiling aims at the quantification of various pre-defined targets. Third, proteomics involves the unbiased identification and quantification of all proteins that exist in certain biological samples taken from an organism grown under defined conditions. Lastly, metabolic profiling involves the segregation and analysis of various metabolites from crude metabolite extracts using mass spectrometry (GCMS/LCMS). However, among the four omics approaches, metabolomics is an advanced and promising approach to studying the metabolic network as it involves the quantification of individual metabolites without any bias. Ideally, proteomic data should exactly characterize the physiological processes taking place at the developmental, genetic or metabolic level. Moreover, the screening of stress-tolerant plants helps to generate novel ideas for enhancing stress tolerance, especially in food crops. Therefore, a comprehensive proteomic approach is of paramount importance to recognize target proteins that are involved to detoxify the HM. HM stress responses in plants have been extensively reviewed by various authors over the last decade (Hossain et al. 2012). Proteomic study helps to unravel the dynamic protein–protein interactions and how the defensive proteins accumulate in the plants upon exposure to HM stress to alleviate the oxidative damage by ROS scavenging. To date, the available proteomic data have revealed a positive correlation between defensive proteins and plant tolerance. To develop a buckwheat proteome reference map, it is necessary to build a buckwheat protein database. To date, only a few buckwheat proteins have been deposited in public databases, namely SwissProt (45 proteins) and Uniprot-KB (466 proteins). Cho et al. (2015) reported that the complete chloroplast genome sequence of Tartary buckwheat has been revealed. In order to unravel the novel metabolites that play a vital role in fighting against abiotic stresses, it is necessary to perform both gel-free and gel-based proteomic analysis coupled with high-resolution mass spectrometry (Figure 4.4). Moreover, these genetic resources can also be used to develop buckwheat varieties to combat abiotic stress. In addition, proteomics analysis also helps to explore potential metabolites, viz. rutin and chiro-inositol, which play a vital role in abiotic stress tolerance in buckwheat (Cho et al. 2015). As per the literature available, there is currently less information that describes the proteomic response of buckwheat to HM stress. It has been reported that the sprouting leaves of 7-day-old common buckwheat turned yellow when subjected to light and dark controlled conditions, demonstrating the restriction of light-dependent protochlorophyllide reductase as revealed during proteomic analysis (Shin et al. 2010). It was also reported that the 13S globulin 3 storage protein (light-inhibited protein) was observed under dark conditions in both buckwheat species. Moreover, another protein (putative pentatricopeptide repeat containing proteins) was also identified in both buckwheat species (leaves and stems) under light-induced conditions, suggesting that various

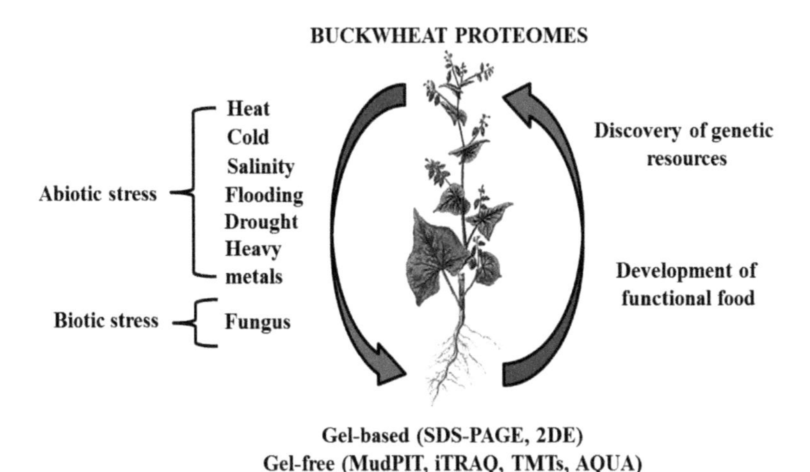

FIGURE 4.4 Potential proteomic application for buckwheat. A gel-based and/or gel-free quantitative proteomic analysis conducted on plants under environmental stresses will contribute to the discovery of genetic resources to improve stress tolerance in buckwheat cultivars or explore its functional components as healthy food supplements. iTRAQ, isobaric tags for relative and absolute quantitation; TMTs, targeted mass tags; MudPIT, multidimensional protein identification technology.

regulator isoforms are involved in the fertility and sterility of buckwheat in a particular light-dependent manner. This type of buckwheat proteomic study would be beneficial to provide an in-depth understanding of buckwheat physiology. It has been recently reported that Tartary buckwheat can withstand high levels of abiotic stress, and this can be attributed to several gene families that play a vital role in signaling pathways, regulating genes and membrane transport. The genomic resources could be ideal repositories for the buckwheat crop-improvement programs (Zhang et al. 2017). Previous reports also revealed abiotic stress tolerance in Tartary buckwheat, but its mechanism is not yet known. It has been found that *FtMYB12* from Tartary buckwheat plays an important role in providing drought and cold resistance by regulating the *COR15A* gene, as reported by Zhou et al. (2015). Reports have also revealed that eight *R2R3-MYB* genes from buckwheat serve a functional role in abiotic stress. Moreover, the buckwheat gene, viz. *FtMYB10*, serves as a negative regulator of salt and drought stress and is linked to abscisic acid (ABA) signaling feedback regulation (Gao et al. 2016). Jeon et al. (2018) investigated the effect of cold stress on the metabolome and transcriptome of Tartary buckwheat and reported several phenylpropanoid biosynthetic transcripts that accumulate in response to cold stress. Data also revealed that most of the phenylpropanoid biosynthetic transcripts get up-regulated in response to cold stress, with the exception of *FtDFR*; moreover, the metabolomics data showed that most of the sugars or sugar-derivatives accumulate in plants subjected to cold stress, while the levels of some amino acids/derivatives were shown to decline in response to cold stress. Similarly, Ma et al. (2019) reported various transcription factors, viz. *NAC*, *bHLH* and *MYB* genes from buckwheat seedlings, that could help to provide salt tolerance. Huang et al. (2018) reported that *FtMYB13* is involved in mediating plant responses to ABA, as well as providing drought and salt stress. He et al. (2019) provide a comprehensive description of the *WRKY* gene family in Tartary buckwheat and report that four *FtWRKYs* (*FtWRKY6*, *74*, *31* and *7*) are related to abiotic stress (salinity, cold, drought and heat stress) and exhibited distinct expression patterns, especially *FtWRKY6*, *74* and *31*, which showed significant up-regulation in response to salinity stress, while *FtWRKY7* showed a response to heat stress; this data provides a basis for screening for more candidate genes for further characterization of *WRKYs* under various stress conditions. Similarly, Li et al. (2019) reported that *FtbZIP83* acts as a positive regulator to salt and drought stress through the ABA signaling pathway composed of *SnRK2-AREB/ABF*. Lv et al. (2019) reported that *FtWRKY46* from Tartary buckwheat acts as a negative regulatory agent in ABA signaling and thus enhances salt tolerance in plants by modulating ROS clearance and stress-related gene expression.

CONCLUSION AND FUTURE POTENTIAL

In conclusion, buckwheat is a potential crop to combat various abiotic stresses and thus could be an ideal candidate for cultivation on marginal lands to improve future food security. The current review concludes that omics technologies have the potential to revolutionize the stress physiology of buckwheat by providing a comprehensive elucidation of the genes/metabolites/proteins or ions that play a pivotal role in abiotic stress tolerance. Moreover, CRISPR/Cas9, the latest genome editing tool, could also be used as a steering technology to engineer buckwheat plants with desired agronomic traits. It is recommended that future experiments must be designed using an integrative omics approach in association with CRISPR/Cas9 technology to develop commercially important engineered plants tolerant to abiotic stress.

REFERENCES

Apel, K. and Hirt, H. 2004. Reactive oxygen species: Metabolism, oxidative stress, and signal transduction. *Annu Rev Plant Biol* 55: 373–399.

Atkinson, N. J. and Urwin, P. E. 2012. The interaction of plant biotic and abiotic stresses: From genes to the field. *J Exp Bot* 63(10): 3523–3544. doi:10.1093/jxb/ers100.

Bohnert, H. J., Gong, Q., Li, P. and Ma, S. 2006. Unraveling abiotic stress tolerance mechanisms – Getting genomics going. *Curr Opin Plant Biol* 9(2): 180–188.

Canter, L. W. 2018. *Environmental Impact of Agricultural Production Activities*. CRC Press, Broken Sound Parkway, NW. doi:10.1201/9781351071796.

Cho, K. S., Yun, B. K., Yoon, Y. H., Hong, S. Y., Mekapogu, M., Kim, K. H. and Yang, T. J. 2015. Complete chloroplast genome sequence of Tartary buckwheat (*Fagopyrum tataricum*) and comparative analysis with common buckwheat (*F. esculentum*). *PLOS ONE* 10(5): e0125332.

Dalcorso, G., Farinati, S. and Furini, A. 2010. Regulatory networks of cadmium stress in plants. *Plant Signal Behav* 5(6): 1–5.

Gao, F., Yao, H., Zhao, H., Zhou, J., Luo, X., Huang, Y., Li, C., Chen, H. and Wu, Q. 2016. Tartary buckwheat FtMYB10 encodes an R2R3-MYB transcription factor that acts as a novel negative regulator of salt and drought response in transgenic Arabidopsis. *Plant Physiol Biochem* 109: 387–396. doi:10.1016/j.plaphy.2016.10.022.

Gygi, S. P., Rist, B., Gerber, S., Turecek, F., Gelb, M. H. and Aebersold, R. 1999. Quantitative analysis of complex protein mixtures using isotope-coded affinity tags. *Nat Biotechnol* 17(10): 994–999.

Hakeem, K. R., Alharby, H. F. and Rehman, R. U. 2019. Antioxidative defense mechanism against lead-induced phytotoxicity in *Fagopyrum kashmirianum*. *Chemosphere* 216: 595–604.

Hall, J. L. and Williams, L. E. 2003. Transition metal transporters in plants. *J Exp Bot* 54(393): 2601–2613. doi:10.1093/jxb/erg303.

Hasegawa, T., Fujimori, S., Havlík, P., Valin, H., Bodirsky, B. L., Doelman, J. C., Fellmann, T., Kyle, P., Koopman, J. F., Lotze-Campen, H., et al. 2018. Risk of increased food insecurity under stringent global climate change mitigation policy. *Nat Clim Chang* 8(8): 699.

He, X., Li, J., Chen, Y., Yang, J. and Chen, X. 2019. Genome-wide analysis of the WRKY gene family and its response to abiotic stress in buckwheat (*Fagopyrum tataricum*). *Open Life Sci* 14(1): 80–96. doi:10.1515/biol-2019-0010.

Hirayama, T. and Shinozaki, K. 2010. Research on plant abiotic stress responses in the post-genome era: Past, present and future. *Plant J* 61(6): 1041–1052.

Hoque, T. S., Burritt, D. J. and Hossain, M. A. 2018. The glyoxalase system: A possible target for production of salinity-tolerant crop plants. In: *Salinity Responses and Tolerance in Plants*, vol. 1. Springer, Switzerland, pp. 257–281. doi:10.1007/978-3-319-75671-4_10.

Horbowicz, M., Kowalczyk, W., Grzesiuki, A. and Mitrus, J. 2011. Uptake of aluminium and basic elements, and accumulation of anthocyanins in seedlings of common buckwheat (*Fagopyrum esculentum* Moench) as a result of increased level of aluminium in nutrient solution. *Ecol Chem Eng* 18: 4.

Horbowicz, M., Debski, H., Wiczkowski, W., Szawara-Nowak, D., Koczkodaj, D., Mitrus, J. and Sytykiewicz, H. 2013. The impact of short term exposure to Pb and Cd on flavonoid composition and seedling growth of common buckwheat cultivars. *Pol J Environ Stud* 22(6): 1723–1730.

Hossain, Z., Hajika, M. and Komatsu, S. 2012. Comparative proteome analysis of high and low cadmium accumulating soybeans under cadmium stress. *Amino Acids* 43(6): 2393– 2416.

Huang, Y., Zhao, H., Gao, F., Yao, P., Deng, R., Li, C., Chen, H. and Wu, Q. 2018. A R2R3-MYB transcription factor gene, *FtMYB13*, from Tartary buckwheat improves salt/drought tolerance in *Arabidopsis*. *Plant Physiol Biochem* 132: 238–248. doi:10.1016/j.plaphy.2018.09.012.

Jeon, J., Kim, J. K., Wu, Q. and Park, S. U. 2018. Effects of cold stress on transcripts and metabolites in Tartary buckwheat (*Fagopyrum tataricum*). *Environ Exp Bot* 155: 488–496. doi:10.1016/j.envexpbot.2018.07.027.

Kumari, A. and Chaudhary, H. K. 2020. Nutraceutical crop buckwheat: A concealed wealth in the lap of Himalayas. *Crit Rev Biotechnol* 40(4): 539–554. doi:10.1080/07388551.2020.1747387.

La-Rocca, N., Andreoli, C., Giacometti, G. M., Rascio, N. and Moro, I. 2009. Responses of the Antarctic microalga *Koliella antartica* (*Trebouxiophyceae*, Chlorophyta) to cadmium contamination. *Photosynthetica* 47(3): 471–479.

Lee, S. H., Ahsan, N., Lee, K. W., Kim, D. H., Lee, D. G., Kwak, S. S., Kwon, S. Y., Kim, T. H. and Lee, B. H. 2007. Simultaneous overexpression of both CuZn superoxide dismutase and ascorbate peroxidase in transgenic tall fescue plants confers increased tolerance to a wide range of abiotic stresses. *J Plant Physiol* 164(12): 1626–1638. doi:10.1016/j.jplph.2007.01.003.

Li, Q., Wu, Q., Wang, A., Lv, B., Dong, Q., Yao, Y., Wu, Q., Zhao, H., Li, C., Chen, H. and Wang, X. 2019. Tartary buckwheat transcription factor FtbZIP83 improves the drought/salt tolerance of *Arabidopsis* via an ABA-mediated pathway. *Plant Physiol Biochem* 144: 312–323. doi:10.1016/j.plaphy.2019.10.003.

Li, X. F., Ma, J. F. and Matsumoto, H. 2002. Aluminium-induced secretion of both citrate and malate in rye. *Plant Soil* 242(2): 235–243.

Lohani, N., Jain, D., Singh, M. B. and Bhalla, P. L. 2020. Engineering multiple abiotic stress tolerance in canola, *Brassica napus*. *Front Plant Sci* 11: 3. doi:10.3389/fpls.2020.00003.

Lv, B., Wu, Q., Wang, A., Li, Q., Dong, Q., Yang, J., Zhao, H., Wang, X., Chen, H. and Li, C. 2019. A WRKY transcription factor, FtWRKY46, from Tartary buckwheat improves salt tolerance in transgenic *Arabidopsis thaliana*. *Plant Physiol Biochem* 147: 43–53. doi:10.1016/j.plaphy.2019.12.004.

Ma, J. F. and Hiradate, S. 2000. Form of aluminium for uptake and translocation in buckwheat (*Fagopyrum esculentum* Moench). *Planta* 211(3): 355–360.

Ma, J. F., Zheng, S. J., Hiradate, S. and Matsumoto, H. 1997. Detoxifying aluminium with buckwheat. *Nature* 390(6660): 569–570.

Ma, W., Kim, J. K., Jia, C., Yin, F., Kim, H. J., Akram, W., Hu, X. and Li, X. 2019. Comparative transcriptome and metabolic profiling analysis of buckwheat (*Fagopyrum tataricum*) Gaertn.) under salinity stress. *Metabolites* 9(10): 225. doi:10.3390/metabo9100225.

Mantri, N., Patade, V., Penna, S., Ford, R. and Pang, E. 2012. Abiotic stress responses in plants: Present and future. In: *Abiotic Stress Responses in Plants.* Springer, New York, pp. 1–19. doi:10.1007/978-1-4614-0634-1_1.

Nadarajah, K. K. 2020. ROS homeostasis in abiotic stress tolerance in plants. *Int J Mol Sci* 21(15): 5208. doi:10.3390/ijms21155208.

Parmar, N., Singh, K. H., Sharma, D., Singh, L., Kumar, P., Nanjundan, J., Khan, Y. J., Chauhan, D. K. and Thakur, A. K. 2017. Genetic engineering strategies for biotic and abiotic stress tolerance and quality enhancement in horticultural crops: A comprehensive review. *3 Biotech* 7(4): 239. doi:10.1007/s13205-017-0870-y.

Pellet, D. M., Grunes, D. L. and Kochian, L. V. 1995. Organic acid exudation as an aluminum-tolerance mechanism in maize (*Zea mays* L.). *Planta* 196(4): 788–795.

Pirzadah, T. B., Malik, B., Tahir, I., Kumar, M., Varma, A. and Rehman, R. U. 2015. Phytoremediation: An eco-friendly green technology for pollution prevention, control and remediation. In: Hakeem, K. R., Sabir, M., Ozturk, M. and Mermut, A. H. (eds.) *Soil Remediation and Plants*. Prospects and Challenges, Elsevier Publications, USA, pp. 107–122.

Pirzadah, T. B., Malik, B., Tahir, I., Qureshi, M. I. and Rehman, R. U. 2018. Characterization of mercury-induced stress biomarkers in *Fagopyrum tataricum* plants. *Int J Phytoremediation* 20(3): 225–236.

Pirzadah, T. B., Malik, B., Tahir, I., Hakeem, K. R., Rehman, R. U. and Alharby, H. F. 2019. Aluminium stress modulates the osmolytes and enzyme defense system in *Fagopyrum* species. *Plant Physiol Biochem* 144: 178–186.

Pirzadah, T. B., Malik, B., Tahir, I., Hakeem, K. R. and Rehman, R. U. 2020. Lead toxicity alters the activities of antioxidant enzymes and modulate the biomarkers in Tartary Buckwheat plants. *Int Biodeterior Biodegrad* 151. doi:10.1016/j.ibiod.2020.104992. http://www.ncbi.nlm.nih.gov/pubmed/104992.

Rascio, N. and Navari-Izzo, F. 2011. Heavy metal hyperaccumulating plants: How and why do they do it? And what makes them so interesting? *Plant Sci* 180(2): 169–181.

Ryan, P. R., Ditomaso, J. M. and Kochian, L. V. 1993. Aluminium toxicity in roots: An investigation of spatial sensitivity and the role of the root cap. *J Exp Bot* 44(2): 437–446.

Shen, R. and Ma, J. F. 2001. Distribution and mobility of aluminium in an Al-accumulation plant, *Fagopyrum esculentum* Moench. *J Exp Bot* 52(361): 1683–1687.

Shin, D. H., Kamal, A. H. M., Suzuki, T., Yun, Y. H., Lee, M. S., Chung, K. Y., Jeong, H. S., Park, C. H., Choi, J. S. and Woo, S. H. 2010. Reference proteome map of buckwheat (*Fagopyrum esculentum* and *Fagopyrum tataricum*) leaf and stem cultured under light or dark. *Aust J Crop Sci* 4(8): 633–641.

Tamura, H., Honda, M., Sato, T. and Kamachi, H. 2005. Pb hyperaccumulation and tolerance in common buckwheat (*Fagopyrum esculentum* Moench). *J Plant Res* 118(5): 355–359.

Vaughan, M. M., Block, A., Christensen, S. A., Allen, L. H. and Schmelz, E. A. 2018. The effects of climate change associated abiotic stresses on maize phytochemical defenses. *Phytochem Rev* 17(1): 37–49. doi:10.1007/s11101-017-9508-2.

Wang, L., Zheng, B., Yuan, Y., Xu, Q. and Chen, P. 2020. Transcriptome profiling of *Fagopyrum tataricum* leaves in response to lead stress. *BMC Plant Biol* 20(1): 54. doi:10.1186/s12870-020-2265-1.

Wang, Y., Wang, L., Zhou, J., Hu, S., Chen, H., Xiang, J., Zhang, Y., Zeng, Y., Shi, Q., Zhu, D. and Zhang, Y. 2019. Research progress on heat stress of rice at flowering stage. *Rice Sci* 26(1): 1–10.

Waqas, M. A., Khan, I., Akhter, M. J., Noor, M. A. and Ashraf, U. 2017. Exogenous application of plant growth regulators (PGRs) induces chilling tolerance in short-duration hybrid maize. *Environ Sci Pollut Res* 24(12): 11459–11471. doi:10.1007/s11356-017-8768-0.

Yang, Z. M., Sivaguru, M., Horts, W. J. and Matsumoto, H. 2001. Aluminium tolerance is achieved by exudation of citric acid from roots of soybean (*Glycine max*). *Physiol Plant* 110(1): 72–74.

Zafar, S. A., Noor, M. A., Waqas, M. A., Wang, X., Shaheen, T. and Raza, M. 2018. Temperature extremes in cotton production and mitigation strategies. In: *Past, Present and Future Trends in Cotton Breeding*. Intech Open, London. doi:10.5772/intechopen.74648.

Zhang, L., Li, X., Ma, B., Gao, Q., Du, H., Han, Y., Li, Y., Cao, Y., Qi, M., Zhu, Y. et al. 2017. The Tartary buckwheat genome provides insights into rutin biosynthesis and abiotic stress tolerance. *Mol Plant* 10(9): 1224–1237.

Zhou, M., Wang, C., Qi, L., Yang, X., Sun, Z., Tang, Y., Tang, Yi, Shao, J. and Wu, Y. 2015. Ectopic expression of *Fagopyrum tataricum* FtMYB12 improves cold tolerance in *Arabidopsis thaliana*. *J Plant Growth Regul* 34(2): 362–371.

Zhu, H., Wang, H., Zhu, Y., Zou, J., Zhao, F. J. and Huang, C. F. 2015. Genome-wide transcriptomic and phylogenetic analyses reveal distinct aluminum-tolerance mechanisms in the aluminum-accumulating species buckwheat (*Fagopyrum tataricum*). *BMC Plant Biol* 15: 16.

Zhu, J. K. 2016. Abiotic stress signaling and responses in plants. *Cell* 167(2): 313–324. doi:10.1016/j.cell.2016.08.029.

Zorb, C., Geilfus, C. M. and Dietz, K. J. 2019. Salinity and crop yield. *Plant Biol* 21(Suppl 1): 31–38. doi:10.1111/plb.12884.

5 Reasons for Declining Buckwheat Production

INTRODUCTION

Buckwheat plays a pivotal role in the functional food sector and is a promising alternative for enhancing the food basket, especially in the higher Himalayan regions, where old generation farmers usually grow buckwheat because of its immense health and nutritional benefits (Park et al. 2019; Sinkovic et al. 2020). However, the cultivation of buckwheat has been in decline over the last two or three decades because of some inherent drawbacks, such as low and unstable yield production, self-incompatibility, flower abortion, seed shattering and lodging (Lee et al. 1996; Farooq et al. 2016; Jacquemart et al. 2012; Dar et al. 2018). In addition, other factors that are responsible for declining buckwheat production include shrinking buckwheat production areas, due to a paradigm shift to newer cash crops that provide farmers better returns; changing food habits; migration; and economic problems (Rana et al. 2012). Moreover, there are certain abiotic stress factors, like temperature stress, photoperiod stress and water stress, that are responsible for low production yield in buckwheat (Delperee et al. 2003; Jacquemart et al. 2012; Dar et al. 2018). It is from this perspective that buckwheat needs global attention, especially from agronomists, molecular breeders and biotechnologists, who need to develop cultivars with improved and desired traits as well as design the best crop-management practices in order to enhance yield and revive buckwheat cultivation on a large scale. In this chapter, we will discuss the inherent problems associated with buckwheat that act as major hindrances to its large-scale production.

MAJOR LIMITATIONS OF THE BUCKWHEAT CROP

The major limitations in buckwheat production are as follows.

Seed Shattering

Seed shattering during the harvesting period is one of the main reasons for a decline in buckwheat production yield, as reported by various authors (Lee et al. 1996; Farooq et al. 2016; Dar et al. 2018). Reports also revealed that seed shattering is more prominent in *F. tataricum* compared to *F. esculentum*, and the non-shattering and shattering forms of buckwheat also showed distinct geographical distribution. It is reported that the Sichuan Province of China, Northern Pakistan and the region from Tibet to Kashmir are dominant in shattering varieties of buckwheat, while the higher Himalayan regions of southern China (1200–3000 a.m. l) are rich in non-shattering varieties (Ohnishi 1998a). It was estimated that the decline in production yield is approximately 40–50% when harvesting is done by mechanical means compared to hand harvesting (Tahir and Farooq 1988; Aufhammer et al. 1994; Radics and Mikohazi 2010). Some studies have revealed that the reason behind buckwheat's tendency to shatter is the weak and brittle pedicle mainly found in the wild cultivar of buckwheat, which is controlled by two complementary dominant genes (*Sht1* and *Sht2*); however, the *Sht1* gene is associated with S-locus, as reported by Matsui et al. (2004). In addition, the formation of an abscission layer across the pedicle is also responsible for seed shattering in buckwheat (Oba et al. 1998; Wang et al. 2005). Furthermore, cultivated buckwheat also showed shattering behavior under adverse conditions because of the small diameter of the pedicle. In addition, the tendency to shatter is lower in tetraploid varieties compared to diploids, although the reason is not yet clear (Hayashi 1992; Fujimura et al. 2001). In a genetic experiment analysis among shattering (*F. tataricum* Potanini) and non-shattering (*F. tataricum* Gaertn) cultivars, it has been reported that the non-shattering lineages possess recessive alleles, which are present at two independently segregating loci and are responsible for affecting the formation of an abscission layer (Fesenko 2006). The shattering gene has been linked to the homostyly locus at a rate of 7.81% and to the allele PS44/Ps42.9 at a rate of 22.54% (Li and Chen 2014). In addition, it exhibited recessive epistasis, i.e., recessive homozygotes at any of the loci suppress the expression of the dominant allele. The data also showed that the recessive allele in the case of *F. esculentum* is present at two or three loci and is responsible for affecting the development of the abscission layer; however, non-shattering varieties only possess a single gene that provides shattering resistance (Wang et al. 2005). However, only two genes have been reported in the case of *F. tataricum*. There are several varieties of buckwheat that possess resistance to shattering, such as the "green flower" mutant (W/SK86GF) and *F. gigantum* (*F. cymosum* × *F. tataricum*), as reported by Suzuki et al. (2012). The shattering resistance in the "green flower" buckwheat cultivar is due to the fact that it has a stronger pedicle than normal buckwheat and is thus regarded as an important genetic resource for crop improvement programs (Alekseeva et al. 1988; Funatsuki et al. 2000) (Figure 5.1).

Flower Abortion

Another limitation of buckwheat crop includes flower abortion, which has a direct effect on its production yield. Though certain barriers have been removed for efficient pollination, some flowers do not produce seeds because of inherent problems like female sterility, self-incompatibility and embryo abortion; these are the main factors in the production of the seed set (Cawoy et al. 2007a, 2007b; Woo et al. 2004). It has been reported that about 76–91% of pollinated flowers do not exhibit any sign of fertilization (Taylor and Obendorf 2001). The seed fails to attain maturity, especially at three distinct levels: immature germination; abnormal flower morphogenesis, which often leads to sterile and undersized gynoecia;

FIGURE 5.1 Picture of flower and seed of W/SK86GF and Kitawasesoba.

and a tendency to indeterminate growth (Farooq et al. 2016). Cawoy et al. (2006) reported that pollen is not regarded as a limiting factor as male sterility seems insignificant, and at mild temperatures, the visibility of pollen is high, at approximately more than 90%. However, female sterility is often considered a major limiting factor responsible for flower abortion, and it might depend upon the location and age of the plant (Bjorkman 1995; Cawoy et al. 2007a). In another study, it has been revealed that flowers within inflorescences do not have an equal chance for seed set, as well as exhibiting a declining trend from the proximal to the distal end in the reproductive structures. Taylor and Obendorf (2001) reported that the first flowers in anthesis have a greater chance of producing seeds compared to later ones. In addition, the size of gynoecia reduces with the passage of time, and many flowers have aberrant undersized pistils (Nagatoma and Adachi 1985; Samborborska-Ciama et al. 1989). It has been estimated that the seed set increases by about 3–10% when all the inflorescences are removed except the terminal one, as reported by Michiyama et al. (1999). In another study, it has been revealed that if half of the inflorescences are removed, compensation mechanisms kick in at the level of the meristems with a 25% increase in flower number and an approximately 10–15% increase in seed set (Halbrecq et al. 2005).

Lodging

Buckwheat lodging, which is characterized by culm dislocation, is another problem, especially in Tartary buckwheat, that has a direct impact on its production yield due to the inhibition of water and nutrient uptake (Pinthus 1973;

Kashiwagi et al. 2005; Zhao and Shang 2009; Murakami et al. 2012). Reports also revealed that it reduces the efficiency of photosynthesis and mechanical harvesting because vascular bundles are damaged during bending or stem breaking in various crops including buckwheat (Hitaka and Kobayashi 1961; Setter et al. 1997; Kashiwagi and Ishimaru 2004; Berry and Spink 2012). However, lodging resistance varies in different cultivars of Tartary buckwheat depending upon their physical properties, lignin content and associated enzymes (Xiang et al. 2019). It has been reported that some cultivars of Tartary buckwheat that possess a thick culm wall or large cross-sectional area exhibited strong lodging resistance (Kashiwagi et al. 2008; Xiang et al. 2014).

Lignin is an essential component of the cell wall that provides structural support to the plant, and it has been reported that lignin shows a positive correlation with the parameters of stem breaking strength/stem puncture strength, demonstrating its prominent role in increasing the strength of stem internodes. Further, it has also been revealed that there is a significant correlation among lignin content and the various enzymes (phenylalanine ammonia-lyase (PAL), tyrosine ammonia-lyase (TAL), 4-coumarate (4CL) and cinnamyl alcohol dehydrogenase (CAD) (Xiang et al. 2019) (Figure 5.2). In another study, Hu et al. (2017) found a significant correlation between lignin content and phenylalanine ammonia-lyase (PAL), 4-coumarate, cinnamyl alcohol dehydrogenase (CAD), which provides lodging resistance in common buckwheat.

OTHER REASONS FOR DECLINING BUCKWHEAT CROP PRODUCTION

Buckwheat crop production has been in decline over the last two or three decades and thus its production area has been shrinking, especially in the higher Himalayan regions. There are several other reasons, and here we have summarized the important ones that have a direct influence on the decline in buckwheat production.

Changing Cropping Pattern

Previously, buckwheat was an integral part of ethnic farming and was cultivated on large scale, especially in hilly regions (Saunders 2010). However, due to the introduction of newer crops and new agricultural policies focusing on cash crops that give more return than the ancient crop that is buckwheat, its cultivation has been on the decline. The area of land given over to buckwheat production has decreased to a large extent in the last two decades, not only in the Himalayan regions but also in other countries, as reported by various authors; however, according to recent reports, buckwheat production increased from 2015 to 2018 due to its immense potential in the functional food sector (Ujihara 1983; Rana et al. 2000; Rao and Pant 2001; Fan and Chan-Kang 2005; Rana et al. 2010a; Zotikov et al. 2010; FAOSTAT 2020) (Figure 5.3).

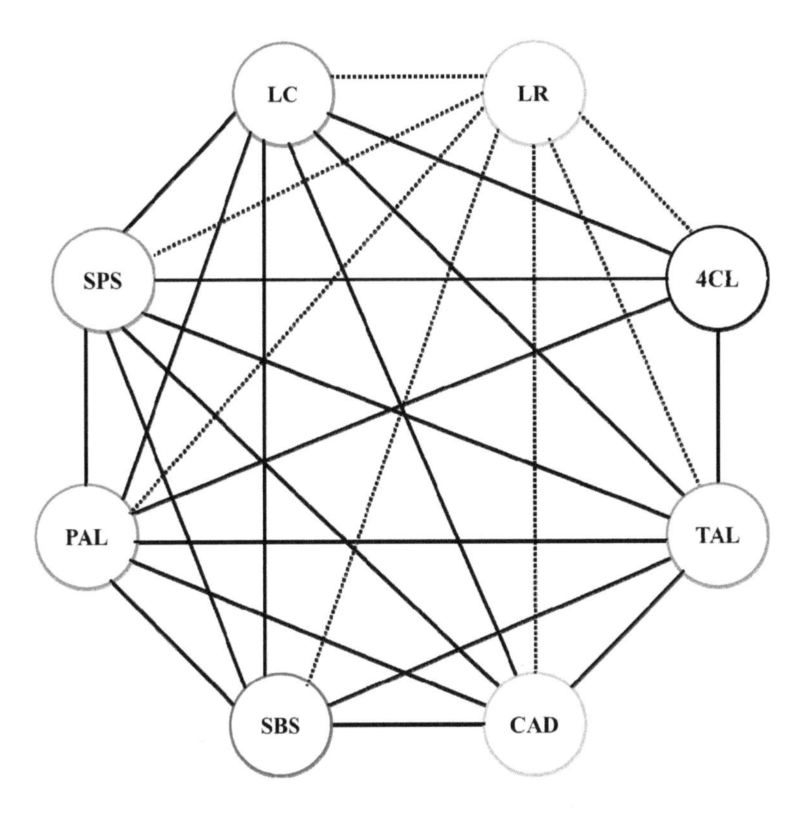

FIGURE 5.2 Network relationships among lignin content, lodging rate, stem mechanical properties, and activities of lignin-related enzymes. (Abbreviations: 4CL-4 – coumarate; CAD – cinnamyl alcohol dehydrogenase; LC – lignin content; LR – lodging rate; PAL – phenylalanine ammonia-lyase; TAL – tyrosine ammonia-lyase; SPS – stem puncture strength; SBS – stem breaking strength.

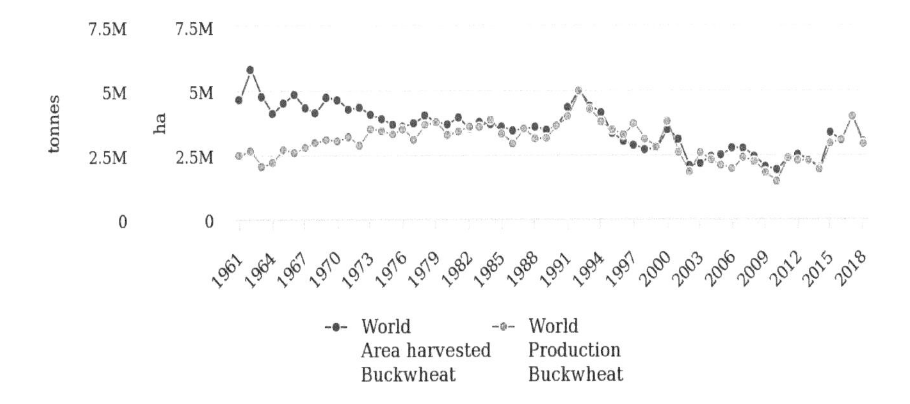

FIGURE 5.3 Graphical representation of world buckwheat production and the total harvested area of buckwheat. (Source: FAOSTAT November 6, 2020.)

Changing Food Habits

Due to changing food habits, there has been a paradigm shift towards the cultivation of modern food crops that are easy to process and give more return in economic terms than the ancient crop buckwheat. With the introduction of novel and hybrid crops, dynamic fluctuations in lifestyle and food habits, a shortage of labor and changing economic conditions, the younger generation has paid less attention to agro-farming, especially the cultivation of ancient crops (Saunders 2010). Moreover, there are also market constraints associated with these ancient crops that have had a direct impact on their cultivation. Furthermore, governments and other organizations have paid less attention to reviving the cultivation of ancient crops, and thus these crops came to be held in low esteem.

ABIOTIC STRESS FACTORS RESPONSIBLE FOR DECLINING BUCKWHEAT PRODUCTION YIELD

Abiotic stress is also one of the important parameters that have a direct impact on the decline in buckwheat production. These abiotic factors include the following.

Temperature Stress

Usually, buckwheat grows well in temperate zones and the optimum temperature for seed germination is approximately 10°C, as reported by Kalinova and Moudry (2003). Moreover, reports also revealed that 1–2 leaf stage is more susceptible to frost conditions because exposure to temperatures between −1 and −3°C for 4–6 hours is fatal; temperature stress also interferes with the reproductive structures, particularly during the flowering period (Bjorkman 1995; Kalinova and Moudry 2003; Sangma and Chrungoo 2010; Jacquemart et al. 2012). The optimal temperature for buckwheat cultivation varies from 18°C to 25°C; however, below 15°C there is delayed initiation and a decline in flower numbers that often leads to the inhibition of flowering or induces early drooping of flowers (Slawinska and Obendorf 2001). Similarly, a high temperature (30°C) and dry wind currents have drastic effects on the production yield by causing flower abortion, drooping of flowers and malformation of the embryo sac, as reported by Gang and Yu (1998). Furthermore, temperature stress and relative humidity also affect the viability of pollen grains. In an experimental analysis, pollen grains lose viability within an hour when placed at 23°C under dry air conditions (Adhikari and Campbell 1998). Michiyama and Sakurai (1999) reported that a temperature of 30°C increases lodging by reducing the stem diameter.

Water stress

Water stress is another parameter that plays an important role in the efficient production of buckwheat (Gang and Yu 1998). As buckwheat possesses a shallow root system, it therefore undergoes withering under drought conditions; however, it quickly rehydrates when an optimal water supply is restored and thus resumes growth (Campbell 1997; Delperee et al. 2003). The production of 1kg of buckwheat seeds requires about 225–315kg of water (Gang and Yu 1998). Reports have also revealed that a limited water supply, especially during the flowering and seed set stages, is detrimental to endosperm development that leads to embryo abortion, therefore reducing crop yield by approximately 50% (Kalinova et al. 2002). Flooding for more than 10 days also constitutes a threat to buckwheat production as it affects the seed weight, lowering flower production (Sugimoto and Sato 2000; Jacquemart et al. 2012). Moreover, due to lodging, the stem comes in close contact with the soil and thus is affected by various pests (Sangma and Chrungoo 2010).

Photoperiod

Buckwheat production is also dependent on the photoperiod as it initiates flowering over a wide range of day length; however, this depends upon the particular cultivar (Nagatoma and Adachi 1985; Hao et al. 1995; Romanova and Koshkin 2010; Hara et al. 2011). Some photosensitive cultivars initiate flowering at a lower node, producing more flowers per inflorescence during short days compared to long days and thus a large ratio of inflorescence (Michiyama et al. 2003). The reason could be due to the interference in the source–sink relationship, causing increasing competition for resources among vegetative and generative organs (Hagiwara et al. 1998).

CONCLUSION AND FUTURE POTENTIAL

Although buckwheat is receiving increased attention in the functional food sector because of its vast array of nutritional properties, it is in decline because of some inherent drawbacks that need to be addressed to revive its cultivation on large scale, especially in the higher Himalayan regions, where it is regarded as an important source of livelihood. Current research and development (R&D) must focus on genetic analysis to identify novel traits like seed-shattering resistance, lodging resistance, determinate growth habit and self-compatibility, which could play an essential role in buckwheat crop improvement programs. Moreover, molecular intervention in this crop has a vast potential that will offer vital clues to the factors involved in its low yield.

REFERENCES

Adhikari, K. N. and Campbell, C. G. 1998. In vitro germination and viability of buckwheat (*Fagopyrum esculentum* Moench) pollen. *Euphytica* 102: 87–92.

Alekseeva, E. S., Malikov, V. G. and Falendysh, L. G. 1988. Green flower form of buckwheat. *Fagopyrum* 8: 79–82.

Aufhammer, W., Fujimoto, F. and Yasue, T. 1994. Development and utilization of the seed yield potential of buckwheat (*F. esculentum*). *Bodenkultur* 45: 37–47.

Berry, P. M. and Spink, J. 2012. Predicting yield losses caused by lodging in wheat. *Field Crops Res* 137: 19–26.

Bjorkman, T. 1995. The effect of pollen load and pollen grain competition on fertilization success and progeny performance in *Fagopyrum esculentum. Euphytica* 83(1): 47–52.

Campbell, C.G. 1997. Buckwheat: *Fagopyrum esculentum* Moench. Promoting the conservation and use of underutilized and neglected crops, 19. Rome: International Plant Genetic Resources Institute.

Cawoy, V., Lutts, S. and Kinet, J. M. 2006. Osmotic stress at seedling stage impairs reproductive development in buckwheat (*Fagopyrum esculentum* Moench). *Physiol Plant* 128(4): 689–700.

Cawoy, V., Halbrecq, B., Jacquemart, A. L., Lutts, S., Kinet, J. M. and Ledent, J. F. 2007a. Genesis of grain yield in buckwheat (*Fagopyrum esculentum* Moench) with a special attention to the seed set. In: Yan, C. and Zongwen, Z. (eds.) *Current Advances in Buckwheat Research: Proceedings of the 10th International Symposium on Buckwheat,* August 14–18. Yangling, China, pp. 111–119.

Cawoy, V., Lutts, S., Ledent, J. F. and Kinet, J. M. 2007b. Resource availability regulates reproductive meristem activity, development of reproductive structures and seed set in buckwheat (*Fagopyrum esculentum* Moench). *Physiol Plant* 131(2): 341–353.

Dar, F. A., Pirzadah, T. B., Malik, B., Tahir, I. and Rehman, R. U. 2018. Molecular genetics of buckwheat and its role in crop improvement. In: Zhou, M., Kreft, I., Tang, Y. and Suvorova, G. (eds). *Buckwheat Germplasm in the World,* 1st edn. Elsevier Publications, USA, pp. 271–286.

Delperee, C., Kinet, J. M. and Lutts, S. 2003. Low irradiance modifies the effect of water stress on survival and growth related parameters during the early developmental stages of buckwheat (*Fagopyrum esculentum* Moench). *Physiol Plant* 119(2): 211–220.

FAO 2020. http://www.fao.org/faostat/en/#data/QC/visualize.

Fan, S. and Chan-Kang, S. 2005. Is small beautiful? Farm size productivity and poverty in Asian agriculture. *Agric Econ* 32(1): 135–146.

Farooq, S., Rehman, R. U., Pirzadah, T. B., Malik, B., Dar, F. A. and Tahir, I. 2016. *Cultivation, Agronomic Practices, and Growth Performance of Buckwheat.* Academic Press, Oxford, pp. 299–320.

Fesenko, I. N. 2006. Non-shattering accessions of *Fagopyrum tataricum* Gaertn carry recessive alleles at two loci affecting development of functional abscission layer. *Fagopyrum* 23: 7–10.

Fujimura, Y., Oba, S. and Horiguchi, T. 2001. Effects of fertilization and poliploidy on grain shedding habit of cultivated buckwheats (*Fagopyrum* spp.). *J Crop Sci* 70(2): 221–225.

Funatsuki, H., Maruyama-Funatsuki, W., Fujino, K. and Agatsuma, M. 2000. Ripening habit of buckwheat. *Crop Sci* 40(4): 1103–1108.

Gang, Z. and Yu, T. 1998. A primary study of increasing the production rate of buckwheat. In: Campbell, C. and Przybylski, R. (eds.), Current Advances in Buckwheat Research. Proceedings of the 7th International Symposium on Buckwheat, Winnipeg, Manitoba, Canada, Aug. 12–14, pp. 18–23.

Hagiwara, M., Inoue, N. and Matano, T. 1998. Variability in the length of flower bud differentiation period of common buckwheat. *Fagopyrum* 15: 55–64.

Halbrecq, B., Romedenne, P. and Ledent, J. F. 2005. Evolution of flowering, ripening and seed set in buckwheat (*Fagopyrum esculentum* Moench): Quantitative analysis. *Eur J Agron* 23(3): 209–224.

Hao, X., Li, G., Yang, W., Zhou, N., Lin, R. and Zhou, M. 1995. The difference and classification of light reaction of buckwheat under different treatments of light duration. In: Matano, T. and Ujihara, A. (eds.), Current Advances in Buckwheat Research. Proceedings of the 6th International Symposium on Buckwheat, Aug. 24–29, Shinshu University Press, Shinshu, Japan, pp. 541–549.

Hara, T., Iwata, H., Okuno, K., Matsui, K. and Obsawa, R. 2011. QTL analysis of photoperiod sensitivity in common buckwheat by using markers for expressed sequence tags and photoperiod-sensitivity candidate genes. *Breed. Sci.* 61: 394–404.

Hayashi, H. 1992. Changes of factors involving in decision of harvest of buckwheat during grain maturation. *The Hokuriku Crop Sci* 28: 81–82.

Hitaka, H. and Kobayashi, H. 1961. Studies on the lodging of rice plants. (II) Source of decreasing yield due to lodging. *Jpn J Crop Sci* 32: 270–276.

Hu, D., Liu, X. B., She, H. Z., Gao, Z., Ruan, R. W., Wu, D. Q. and Yi, Z. L. 2017. The lignin synthesis related genes and lodging resistance of *Fagopyrum esculentum. Biol Plant* 1: 138–146.

Jacquemart, A. L., Cawoy, V., Kinet, J. M., Ledent, J. F. and Quinet, M. 2012. Is buckwheat (*Fagopyrum esculentum* Moench) still a valuable crop today. *Eur J Plant Sci Biotechnol* 6(2): 1–10.

Kalinova, J. and Moudry, J. 2003. Evaluation of frost resistance in varities of common buckwheat (*Fagopyrum esculentum* Moench). *Plant Soil Environ* 49: 410–413.

Kalinova, J., Moudry, J. and Curn V. 2002. Technological quality of common Buckwheat (*Fagopyrum esculentum* Moench). *Rostlinna Vyroba* 48: 279–284.

Kashiwagi, T. and Ishimaru, K. 2004. Identification and functional analysis of a locus for improvement of lodging resistance in rice. *Plant Physiol* 134(2): 676–683.

Kashiwagi, T., Sasaki, K. and Ishimaru, K. 2005. Factors responsible for decreasing sturdiness of the lower part in lodging of rice (*Oryza sativa* L. *Plant Prod Sci* 2: 166–172.

Kashiwagi, T., Togawa, E., Hirotsu, N. and Ishimaru, K. 2008. Improvement of lodging resistance with QTLs for stem diameter in rice (*Oryza sativa* L.). *Theor Appl Genet* 117(5): 749–757.

Lee, J. H., Aufhammer, W. and Kubler, E. 1996. Produced, harvested and utilizable grain yield of the pseudocereals buckwheat (*Fagopyrum esculentum*, Moench), quinoa (Chenopodium quinoa, Wild) and amaranth (*Amaranthus hypochondriacus*, L × *A. hybridus*, L.) as affected by production techniques. *Bodenkultur* 47: 5–14.

Li, J. H. and Chen, Q. F. 2014. Inheritance of seed protein subunits of common buckwheat (*Fagopyrum esculentum* Moench) cultivar Sobano and its Homostylous wild type. *J Agric Sci* 6(6): 1–9.

Matsui, K., Nishio, T. and Tetsuka, T. 2004. Genes outside the S supergene suppress S functions in buckwheat (Fagopyrum esculentum). *Ann Bot* 94(6): 805–809.

Michiyama, H., and Sakurai, S. 1999. Effect of day and night temperatures on the growth and development of common buckwheat (*Fagopyrum esculentum* Moench). *Jpn J Crop Sci.* 68: 401–407.

Michiyama, H., Arikuni, M., Hirano, T. and Hayashi, H. 2003. Influence of day light before and after the start of anthesis on the growth, flowering and seed setting in common buckwheat (*Fagopyrum esculentum* Moench). *Plant Prod. Sci.* 6: 235–242.

Michiyama, H., Tachimoto, A. and Hayashi, H. 1999. Effect of defloration and restriction of the number of flower clusters on the progression of successive flowering and seed-setting in common buckwheat (*Fagopyrum esculentum* Moench). *Jpn J Crop Sci* 68(1): 91–94.

Murakami, T., Yui, M. and Amaha, K. 2012. Canopy height measurement by photogrammetric analysis of aerial images: Application to buckwheat (*Fagopyrum esculentum* Moench) lodging evaluation. *Comput Electron Agric* 89: 70–75.

Nagatoma, T. and Adachi, T. 1985. *Fagopyrum esculentum*. In: Halevy, A. H. (ed.) *Handbook of Flowering*. Vol III. CRC Press, Boca Raton, FL, pp. 1–8.

Oba, S., Suzuki, Y. and Fujimoto, F. 1998. Breaking strength of pedicels and grain shattering habit in two species of buckwheat (*Fagopyrum* spp.). *Plant Prod Sci* 1(1): 62–66.

Ohnishi, O. 1998a. Search for the wild ancestor of common buckwheat. Description of new *Fagopyrum* (*Polygonaceae*) species and their distribution in China and Himalayan hills. *Fagopyrum* 15: 18–28.

Park, B. I., Kim, J., Lee, K., Lim, T. and Hwang, K. T. 2019. Flavonoids in common and Tartary buckwheat hull extracts and antioxidant activity of the extracts against lipids in mayonnaise. *J Food Sci Technol* 56(5): 2712–2720.

Pinthus, M. J. 1973. Lodging in wheat, barley, and oats: The phenomenon, its causes, and preventive measures. *Adv Agron* 25: 209–263.

Radics, L. and Mikohazi, D. 2010. Principles of common buckwheat production. In: Dobranszki, J. (ed.), Buckwheat 2. *Eur Plant J Sci Biotechnol* 4(1): 57–63.

Rana, J. C., Sharma, B. D. and Gautam, P. L. 2000. Agri-diversity erosion in the northwest Indian Himalayas – Some case studies. *Ind J Plant Genet Res* 13: 252–258.

Rana, J. C., Singh, A., Sharma, Y., Pradheep, K. and Mendiratta, N. 2010. Dynamics of plant bioresources in western Himalayan Region of India – Watershed based case study. *Curr Sci* 98: 192–203.

Rana, J. C., Chauhan, R. C., Sharma, T. R. and Gupta, N. 2012. Analyzing problems and prospects of buckwheat cultivation in India. *Eur J Plant Sci Biotechnol* 6(2): 50–56.

Rao, K. S. and Pant, R. 2001. Land use dynamics and landscape change pattern in a typical micro watershed in the mid elevation zone of central Himalaya, India. *Agric Ecosys Environ* 86(2): 113–123.

Romanova, O. and Koshkin, V. 2010. Photoperiod response of landraces and improved varieties of buckwheat from Russia and from the main buckwheat cultivating countries. In: Dobranszki, J. (ed.), Buckwheat 2. *Eur. J. Plant Sci. Biotechnol.* 4(1): 123–127.

Samborborska-Ciana, A., Januszewicz, E. and Ojczyk, T. 1989. The morphology of buckwheat flowers depending on the course of plant flowering. *Fagopyrum* 9: 23–26.

Sangma, S. C. and Chrungoo, N. K. 2010. Buckwheat gene pool: potentialities and drawbacks for use in crop improvement programmes. In: Dobranszki, J. (ed.), *Buckwheat 2. Eur. J. Plant Sci. Biotechnol.* 4(1): 45–50.

Saunders, M. 2010. Losing ground: An uncertain future for buckwheat farming in its center of origin. In: Zotikov, V. I. and Parakhin, N. V. (eds) *Advances in Buckwheat Research: Proceedings of the 11th International Symposium on Buckwheat.* Orel, Russia, pp. 60–68.

Setter, T. L., Laureles, E. V. and Mazaredo, A. M. 1997. Lodging reduces yield of rice by self-shading and reductions in canopy photosynthesis. *Field Crops Res* 49(2–3): 95–106.

Sinkovic, L., Kokalj, D., Vidrih, R. and Meglic, V. 2020. Milling fractions fatty acid composition of common (*Fagopyrum esculentum* Moench) and Tartary (*Fagopyrum tataricum* (L.) Gaertn) buckwheat. *J Stored Prod Res* 85, http://www.ncbi.nlm.nih.gov/pubmed/101551.

Slawinska, J. and Obendorf, R. L. 2001. Buckwheat seed set in planta and during in vitro inflorescence culture: evaluation of temperature and water deficit stress. *Seed Sci. Res.* 11: 223–233.

Sugimoto, H. and Sato, T., 2000. Effects of excessive soil moisture at different growth stages on seed yield of summer buckwheat. *Jpn. J. Crop Sci.* 69: 189–193.

Suzuki, T., Mukasa, Y., Morishita, T., Takigawa, S. and Noda, T. 2012. Traits of shattering resistant buckwheat "W/SK86GF". *Breed Sci* 62(4): 360–364.

Tahir, I. and Farooq, S. 1988. Review article on buckwheat. Buckwheat Newsletter, *Fagopyrum* 8: 33–53.

Taylor, D. P. and Obendorf, R. L. 2001. Quantitative assessment of some factors limiting seed set in buckwheat. *Crop Sci* 41(6): 1792–1799.

Ujihara, A. 1983. *Studies on the Ecological Features and the Potentials as Breeding Materials of Asian Common Buckwheat Varieties (Fagopyrum esculentum M.)* (PhD thesis). Faculty of Agriculture, Kyoto University, Japan, pp. 48–56.

Wang, Y., Scarth, R. and Campbell, G. C. 2005. Inheritance of seed shattering in interspecific hybrids between *Fagopyrum esculentum* and *F. homotropicum*. *Crop Sci* 45(2): 693–697.

Woo, S. H., Omoto, T., Kim, H. S., Park, C. R., Campbell, C., Adachi, T. and Jong, S. K. 2004. Breeding improvement of processing buckwheat: Prospects and problems of interspecific hybridation. In: Faberova, I., Dvoracek, V., Cepkova, P., Hon, I., Holubec, V. and Stehno, Z. (eds.) *Advances in Buckwheat Research (I): Proceedings of the 9th International Symposium on Buckwheat*, August 18–22. Prague. *Czech Republic* 35Q: 354.

Xiang, D. B., Li, J., Fan, Y., Peng, L. X., Song, C., Zhao, G. and Zhao, J. L. 2014. The effect of planting density on lodging resistance and yield of Tartary buckwheat. *Chin Agric Sci Bull* 6: 242–247.

Xiang, D., Song, Y., Wu, Q., Ma, C., Zhao, J., Wan, Y. and Zhao, G. 2019. Relationship between stem characteristics and lodging resistance of Tartary buckwheat (*Fagopyrum tataricum*). *Plant Prod Sci* 22(2): 202–210.

Zhao, G. and Shang, F. 2009. *Tartary Buckwheat of China*. Science Press, Beijing.

Zotikov, V. I., Sidorenko, V. S. and Tsukanova, Z. R. 2010. Problems and perspectives of development of seed farming of groats crops (buckwheat, millet) in Russia. In: Zotikov, V. I. and Parakhin, N. V. (eds.) Advances in Buckwheat Research: Proceedings of the 11th International Symposium on Buckwheat, July 19–23. Orel, Russia, pp. 381–388.

6 Strategies for Buckwheat Crop Improvement Using Modern Approaches

INTRODUCTION

Although agricultural productivity experienced a significant increase in the second half of the 20th century, only 12 crops provide 75% of the world's food supply, and 50% of the energy requirement is just met by the three main crops: rice, maize and wheat (FAO 2017; Allen and de Brauw 2018). The narrowness of the agriculture sector raises certain questions as to how major crops are going to alleviate poverty and combat future food insecurity. As a result, rural incomes and the future supply of food are at high risk because of the decline in production yield and dependence on a handful of crops. The decrease in the portfolio of species used in agriculture also decreases the ability of farmers and ecosystems to adapt to new environments, needs and opportunities (Joshi et al. 2019; Pirzadah et al. 2020). To highlight these challenges, it is necessary to widen the scope of research to include those neglected crops that will not only address the issue of future food security but will also tackle the hidden hunger crisis. In this context, various research organizations around the globe are focusing on the acceleration and domestication of these neglected or underutilized or orphan crops. Therefore, in addressing such severe issues, the scientific community should broaden its scope and focus on research and development (R&D) programs pertaining to the exploitation of these underutilized crops in order to enhance the food basket. These crops must also have the potential to achieve a high market value due to their product characteristics. In this context, the International Centre for Underutilized Crops (ICUC) has identified buckwheat as a crop with strong potential for domestication. Buckwheat (*Fagopyrum* spp.) is an ancient crop that has long been grown in East Asia and the Himalayan region (Kreft 2016; FAOSTAT 2018). The renewed interest in buckwheat is based on its high nutritional profile and qualities such as its high protein/amino acid content, its flavonoid content, its gluten-free nature and the fact that it requires minimum input due to its ability to grow on marginal lands. UNESCO has placed the buckwheat crop under the category of important crops because of its declining cultivation and exploitation in the wild (Small 2017). In spite of its high demand in the functional food sector, the area of land devoted to the production and cultivation of buckwheat globally has been on the decline

TABLE 6.1
Sources of Useful Traits Identified in Buckwheat Germplasm

Trait of interest	Value	Species	References
Days to flowering	<40 days	*F. tataricum*	Chauhan et al. (2010)
Days to maturity	<80 days	*F. esculentum, F. tataricum*	Rana et al. (2016), Rauf et al. (2020)
Seed yield per plant	>100 g	*F. tataricum*	Rana et al. (2016)
1000-seed weight	>25 g	*F. esculentum, F. tataricum*	Chauhan et al. (2010), Rauf et al. (2020)
Lysine content	>4.50%	*F. tataricum*	Zheng et al. (2011), Rana et al. (2016)
Total phenols	>1.60%	*F. cymosum, F. tataricum*	Rana and Sharma (2000), Zhao et al. (2002)
Total proteins	>14%	*F. cymosum, F. esculentum*	Przybylski and Gruczynska (2009), Tang et al. (2011)
Quercetin content	>0.2 l g/mg	*F. esculentum, F. tataricum*	Yu et al. (2019), Rauf et al. (2020)
Rutin content	>17 l g/mg	*F. tataricum*	Gupta et al. (2012), Yu et al. (2019)

over the last five decades in comparison to cereal crops, and the reason may be associated with the lack of R&D on buckwheat crop improvement programs and the introduction of newer crop species. Significant genetic improvement of buckwheat using conventional breeding has not been accomplished because of some drawbacks associated with buckwheat, viz. its indeterminate growth habit, flower abortion, lodging, self-incompatibility and low yield (Neskovic et al. 1995; Woo et al. 1998; Dar et al. 2018). Some of the useful traits that have been identified in the buckwheat germplasm are depicted in Table 6.1. From this perspective, the integration of advanced biotechnological and omics approaches along with classical breeding tools is the key to the genetic improvement of the buckwheat crop. Recently, considerable progress at the genetic level has been made in buckwheat that could help to predict it as a future golden crop of the cropping system. This chapter summarizes the strategies for buckwheat crop improvement using modern approaches like biotechnological and genomic advances. Besides, we also offer brief insights into emerging molecular approaches to revive and accelerate this traditional crop, which has immense agricultural potential.

VARIOUS PRODUCTION CONSTRAINTS AND BREEDING TARGETS

For buckwheat crop improvement, the main objectives of the breeding program should focus on enhanced yield production, determinate growth habit, reduced flower abortion, lodging resistance and synchronized flowering, as reported by various authors (Campbell 1997; Song et al. 2019; Morishita et al. 2020). It has been reported that a lack of photosynthates, especially during the grain filling stage, is one of the prime factors responsible for low yield in buckwheat (Yang et al. 1998). Reports have revealed that buckwheat requires only 24% of the blossoms which are being produced to improve the seed yield two-fold. Thus breeding buckwheat cultivars possessing optimum flowers could be a great approach to switching the nutrients towards the grain filling stage that are otherwise being exhausted for the formation of blooms. Another important breeding objective associated with Tartary buckwheat involves an easy dehulling process, an essential parameter during the use of post-harvesting technology. Campbell (1997) reported the local landraces, viz. rice buckwheat with an easy dehulling process, could be an important repository for transferring their traits to high-yielding buckwheat cultivars. Moreover, other important breeding targets involve early maturity, enhanced groat percentage, the presence of zero allergic protein, lodging resistance and determinate growth habit. Frost susceptibility in common buckwheat is another hindrance that has a major impact on the production yield of buckwheat (Zhou et al. 2016). However, there are some cultivars of Tartary buckwheat that possess tolerance to frost conditions, and thus such frost-tolerant cultivars could be used as an important source of desirable genes for bioprospecting (Campbell 1995; Zhou et al. 2016). Lodging is another problem associated with buckwheat that directly affects production yield and thus has economic consequences, and therefore determining how to improve stem quality, especially mechanical strength, has become the main target for increasing crop lodging resistance and ultimately production yield. Elucidating how plant density regulates lignin biosynthesis and its relationship with lodging resistance in buckwheat is of great economic importance. Schematic representation of the bending moment (BM) will be measured using the equation BM = SL (stem length) × FW (fresh weight). Then, lodging index (bending moment/breaking resistance × 100) will be measured for N3 internode as reported by Amano et al. (1993) (Figure 6.1). Further, the important traits responsible for improving nutraceutical properties and the knockdown of genes responsible for self-incompatibility and allergens need to be included as prime targets in the various breeding programs so that the buckwheat crop becomes more competitive with main cereal crops. Due to the easy emasculation and self-pollinating nature of Tartary buckwheat, it could be an important source for retrieving transgressive segregants usually for rutin concentration using a selective breeding technique. For this purpose, different molecular markers, viz. simple sequence repeats (SSRs), expressed sequence tag (ESTs), amplified fragment length polymorphism (AFLP) and array based markers have been developed (Konishi and Ohnishi 2006; Hara et al. 2011; Yasui et al. 2004; Yabe et al. 2014).

HINDRANCES TO THE BUCKWHEAT BREEDING PROGRAM

Due to the presence of self-incompatibility and the dimorphic nature of its flowers, sweet buckwheat is considered to be an allogamous crop that hinders the breeding of pure line cultivars and thus fixes the economic traits (Figure 6.2). Moreover, the emasculation process is considered an inconvenient process because of the complicated structure of the floral organs (Yasui et al. 2016). Marshall (1969)

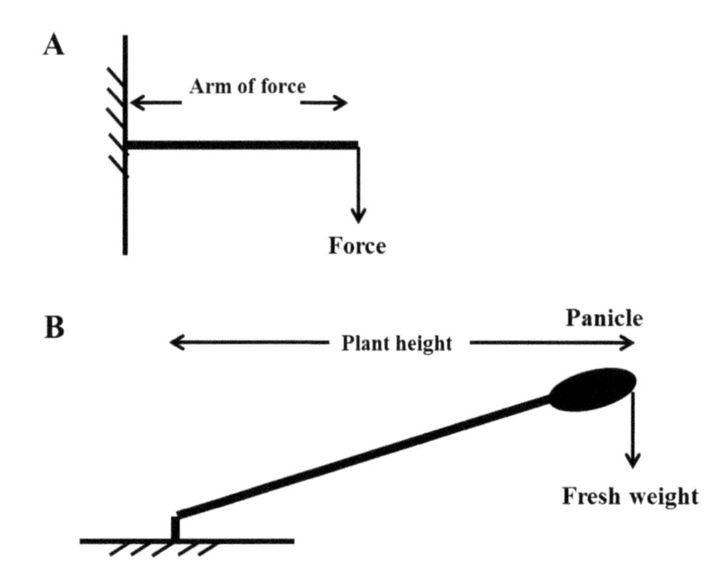

FIGURE 6.1 Schematic representation of bending moment (BM) will be calculated using the following formula: BM = SL (stem length) × FW (fresh weight).

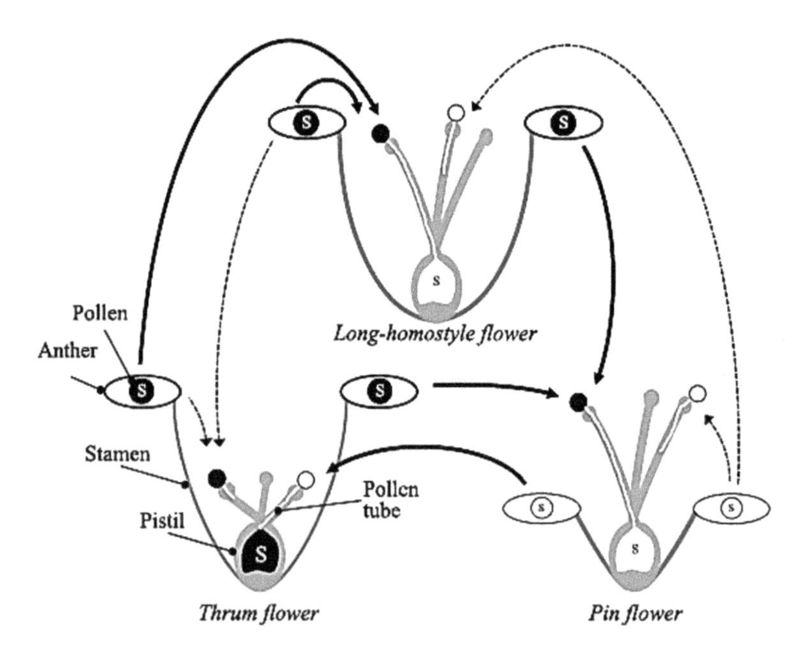

FIGURE 6.2 Diagram of cross-compatibility/incompatibility among pin, thrum and long-homostyle plants. Buckwheat is a sporophytic SI species, and the SI phenotype of the pollen is determined by the genotype of its diploid parent. Solid arrows, compatible crosses; dashed arrows, incompatible crosses.

reported that some cultivars of sweet buckwheat are self-compatible and homomorphic in nature and possess the ability to self-fertilize. However, their application in breeding programs is limited because of serious inbreeding depression. This could perhaps be due to the number of harmful homozygous (ss) recessive mutants that generally remain dormant when in heterozygous (Ss) conditions. Contrary to sweet buckwheat, bitter buckwheat is considered a self-compatible autogamous species that makes it easier to develop large segregating progenies and thus preserve the genetic purity of progressive breeding lines (Woo et al. 2010). In order to overcome the breeding hindrances associated with buckwheat, it is necessary to use effective and potential hybridization tools to cross elite lines for advancing the genetic base of cultivars.

BREAKTHROUGHS IN BUCKWHEAT BREEDING STRATEGIES

The various breakthroughs achieved in buckwheat breeding programs are depicted in Table 6.2. The first is the release of the buckwheat variety "Bogatyr" in 1938, which was developed using mass selection from the Russian landrace population (Suvorova and Zhou 2018). There have been reports regarding interspecific hybridization among *Fagopyrum esculentum* and *F. cymosum* to improve the buckwheat crop, but the sterility of the resultant hybrid acts as a major hindrance to its success (Ujihara et al. 1990; Suvorova et al. 1994; Hirose et al. 1995a; Woo et al. 1999a; Asaduzzaman et al. 2009) (Table 6.3). Krotov and Golubeva (1973) revealed that *F. giganteum* was first reported as interspecific hybrid that was developed by crossing *F. cymosum* and *F. tataricum*. Similarly, Campbell (1995) developed self-pollinating buckwheat by crossing *F. esculentum* and *F. homotropicum* in what was considered a major breakthrough in the buckwheat breeding program. Furthermore, several experiments have been conducted to improve the success rate of interspecific hybridization among these two cultivated buckwheat species (Samimy et al. 1996; Wang et al. 2002; Niroula et al. 2006; Azaduzzaman et al. 2009). It is reported that interspecific hybridization among cultivated buckwheat species is a promising route to improve production yield. The various obstacles involved in fertility should be analyzed, and novel biotechnological tools should be employed to combat those obstacles (Hirose et al. 1995b; Samimy et al.1996; Wang et al. 2002; Niroula et al. 2006). The restoration of fertile diploid and tetraploid hybrids among *F. tataricum* × *F. esculentum* by means of ovule culture is one of the crucial steps in the interspecific hybridization that leads to the formation of the F2 generation, which are more fertile compared to their parents or the F1 generation, and this successful attempt could provide a breakthrough in buckwheat crop improvement (Azaduzzaman et al. 2009). Samimy et al. (1996) carried out one such experiment to develop a hybrid using a polyethylene glycol (PEG) mediated fusion of mesophyll protoplast of *F. esculentum* and hypocotyl protoplast of *F. tataricum*. Furthermore, a successful breakthrough was achieved in Russia and Eastern Europe in developing a cultivar with a determinate growth habit using recessive allele (*det*), and it is estimated that this type of cultivar would enhance the average yield of buckwheat to almost

TABLE 6.2
Various Breakthroughs in Buckwheat Breeding Programs

Study/problem	Organ used	Result	References
Anther culture	Anther	Induction of haploid or diploid plants	Bohanec et al. (1993)
Interspecific hybridization using embryo rescue	Embryo	Successful production of interspecific hybrids from crosses between *F. cymosum* × *F. esculentum* and *F. esculentum* × *F. tataricum*	Woo et al. (1995), Woo et al. (2002), Niroula et al. (2006)
Gametophyte selection	Pollen	Pollen competition influenced genetic structure and vigor of buckwheat mapping population	Bjorkman (1995)
Self-incompatibility in interspecific crosses	Pollen	Occurrence of unilateral incompatibility and dimorphic self-incompatibility	Hirose et al. (1995a)
In vitro germination and viability	Pollen	Temperature and flower age–influenced pollen longevity	Adhikari and Campbell (1998)
Inflorescence	Pollen	Light-affected photosynthetic rate and flowering time	Quinet et al. (2004)
Cytogenetics	Pollen	Crossing auto-tetraploid plants with diploid parent produced auto-triploid and trisomic progenies	Chen et al. (2007)
Embryo development in interspecific crosses	Pollen	Highly compatible pollination occurred between *F. esculentum* × *F. cymosum* and *F. esculentum* × *F. homotropicum*	Woo et al. (2008)
Enhanced seed development by backcross	Pollen	Desirable agronomic traits from wild *F. homotropicum* transferred to cultivated *F. esculentum*	Shin et al. (2009)

double (Fesenko et al. (2006, 2016). Chen et al. (2018) reported a tetraploid variety, namely "Hongxin Jinqiao," with less seed shattering, which was developed using accessions of *F. cymosum* that possess a reduced seed-shattering habit.

ROLE OF GENOMICS IN BUCKWHEAT CROP IMPROVEMENT

Buckwheat, being an important functional food, has attracted the attention of scientists worldwide, who have focused on the desirable and undesirable traits associated with buckwheat. Over the last two decades, employing advancements

TABLE 6.3
Status of Interspecific Hybridization between the Species of Large Achene *Cymosum* Group of Genus *Fagopyrum*

Interspecific cross	Method used	Result	Reference
F. esculentum (2×) × *F. cymosum* (2×)	Conventional crossing	Embryos of early globular stage only	Shaikh et al. (2002)
F. esculentum (4×) × *F. cymosum* (4×)	Ovule rescue technique	Hybrids (4×) with perennial growth habit	Suvorova (2001)
F. esculentum (2×) × *F. cymosum* (2×)	Ovule rescue technique	Hybrids (2×) characterized by perennial growth habit and heterostyly flower	Hirose et al. (1995b)
F. esculentum (4×) × *F. cymosum* (4×)	Embryo rescue	Hybrids (4×) capable of self-fertilization	Rumyantseva et al. (1995
F. esculentum (2×) × *F. cymosum* (2×)	Embryo rescue	Vigorous but self-sterile hybrids (2×)	Woo et al. (1999b)
[(*F. esculentum* × *F. cymosum*) × *F. homotopicum*] × *F. homotopicum*	Multiple crossing and ovule rescue	Vigorous but sterile tri-species hybrids	Suvorova (2010)
F. tataricum (4×) × *F.cymosum* (4×)	Conventional crossing	Amphidiploid fertile tetraploid (2n = 4× = 32) hybrid. Later designated as a separate species, *F. giganteum*	Krotov and Golubeva (1973)
F. tataricum (4×) × *F. giganteum* (4×)	Conventional crossing	Early maturing, self-fertilizing amphidiploid hybrids (4×). Designated as a new man-made species, *F. hybridum*	Fesenko and Fesenko (2010)
F. tataricum × *F. esculentum*	Ovule rescue technique	Sterile hybrids (2× and 4×)	Samimy et al. (1996) Hirose et al. (1995b), Wang et al. (2002), Niroula et al. (2006)
F. tataricum × *F. esculentum*	Ovule rescue technique	Fertile hybrids (2× and 4×). The fertile hybrids produced F1 and F2 generation	Azaduzzaman et al. (2009)
F. esculentum (4×) × *F. homotropicum* (4×)	Embryo rescue	Fully fertile hybrids (4×)	Campbell (1995)
F. esculentum (4×) × *F. homotropicum* (4×)	Conventional crossing	Fertile hybrids (2× and 4×)	Fesenko et al. (2001), Kim et al. (2002)

in biotechnology, there has been a constant effort to tackle the issues and challenges of buckwheat crop improvement. Currently, we have well-assembled reference genomes, especially for those cultivated buckwheat species, viz. sweet buckwheat (size: 1.2 Gbp) and bitter buckwheat (size: 0.48 Gbp), that are considered an excellent source for the genome-assisted breeding program (Yasui et al. 2016; Zhang et al. 2017). Moreover, reports have also revealed the availability of two chloroplast genome sequences associated with buckwheat, and these reference genomes play an important role in the recognition of new candidate genes prognosticated to be associated with self-incompatibility, the biosynthesis of rutin and abiotic stress tolerance (Liu et al. 2016; Wang et al. 2017). Various genomic markers have been involved in exploring the diversity and evolutionary studies of buckwheat. However, for the rapid development of molecular markers in buckwheat, expressed sequence tags (ESTs) play a vital role (Chauhan et al. 2010; Rana et al. 2016). These authors identified the genes involved in the biosynthesis of rutin in sweet and bitter buckwheat using a comparative genomic approach. Table 6.4 summarizes the information regarding marker-assisted selection in buckwheat. Chauhan et al. (2010) reported that genetic maps are considered another important genomic resource for carrying out rigorous analysis of the association among quantitative trait loci (QTLs) and agronomic traits for marker-assisted selection. These genetic maps play an important role in the identification of QTLs that govern photoperiod sensitivity and stem length, especially in cultivated species of buckwheat. However, limited research has been conducted in developing genetic mapping populations of buckwheat in terms of rapid improvements made in other crops associated with the production of multiple advanced generation intercross (MAGIC), nested associated mapping (NAM) and recombinant inbred lines (RILs) populations (Yasui et al. 2004; Pan and Chen 2010; Hara et al. 2011; Xiaolei et al. 2013; Yabe et al. 2014). Recently, another approach that involves speed breeding has been employed for the rapid development of cultivars especially in the cereal crops to govern six crop cycles/year compared to one or two cycles, as reported by Watson et al. (2018). As buckwheat is a short duration crop (3–4 months), this approach could be utilized to improve the production yield of buckwheat by achieving six crop cycles/annum (Joshi et al. 2019). The integration of a genomics-assisted breeding approach is an immediate step towards improving buckwheat production. Even though the potential of genomic selection to complement mass selection in buckwheat has been established, the recently developed genome-editing tool CRISPR/Cas9 technology will be helpful to combat the inherent problems associated with buckwheat and thus revolutionize the functional food sector and yield production of buckwheat (Yabe et al. 2018). Moreover, the capability of genomic selection in buckwheat has recently been established, and the factual results of the pilot project showed that genomic prediction could accelerate the genetic gains for nutritional traits/generation through early selection and possesses an immense capability for the biofortification breeding of buckwheat (Yabe et al. 2018) (Figure 6.3). The Genome-Wide Association Study (GWAS) is another efficient genomic-based breeding protocol that has the potential to

TABLE 6.4
Progress of Marker-Assisted Selection in Buckwheat

Study	Marker	Result	References
Population genetics of *F. esculentum*	RAPD	Construction of phylogenetic trees for landraces from all over the world	Murai and Ohnishi (1996)
Identification of molecular markers linked to the homostylar (Ho) gene	RAPD	Generation of F2 population from an interspecific cross between *F. esculentum* and *F. homotropicum*	Ali et al. (1998)
Origin of *F. tataricum*	RAPD	Construction of phylogenetic tree based on RAPD markers	Tsuji and Ohnishi (2000)
PCR-based DNA fingerprinting	RAPD	Uncovering of species relationship between 28 different accessions belonging to 14 different species	Sharma and Jana (2002)
Characterization of interspecific hybrids between *F. esculentum* and *F. tataricum*	RAPD	Production of hybrids from interspecific crosses using ovule rescue method	Asaduzzaman et al. (2009)
Genetic mapping of F. esculentum	RAPD	Construction of linkage map involving RAPD markers	Pan and Chen (2010)
Conversion of AFLP marker to a simple PCR-based marker	AFLP	Identification of tightly linked markers associated with cell compatibility	Nagano et al. (2001)
Identification of AFLP markers linked to non-seed shattering (*Sht 1*) locus	AFLP	Identification of 5 AFLP markers linked to *Sht 1* locus	Matsui et al. (2004)
AFLP linkage analysis of cultivated F. esculentum and wild F. homotropicum	SSR	Development of interspecific linkage map and SSR markers	Yasui et al. (2004)
Development of SSR markers	SSR	Development of 136 SSR markers in *F. esculentum*	Ma et al. (2009)
Assessment of genetic diversity and population structure	SSR	Low genetic differentiation due to out-crossing and self-incompatibility	Song et al. (2011)

identify the regions related to the target traits. GWAS has not been reported yet for buckwheat, probably due to the lack of long-range linkage information like the pseudo-molecules of chromosomes that will play a vital role in identifying the agronomically desired and beneficial traits in buckwheat. GWAS helps to figure out the phenotypes of the target allele, especially in the self-compatible lines possessing the alleles in the homozygous state at each locus (Korte and Farlow 2013; Huang and Han 2014; Matsui and Yasui 2020).

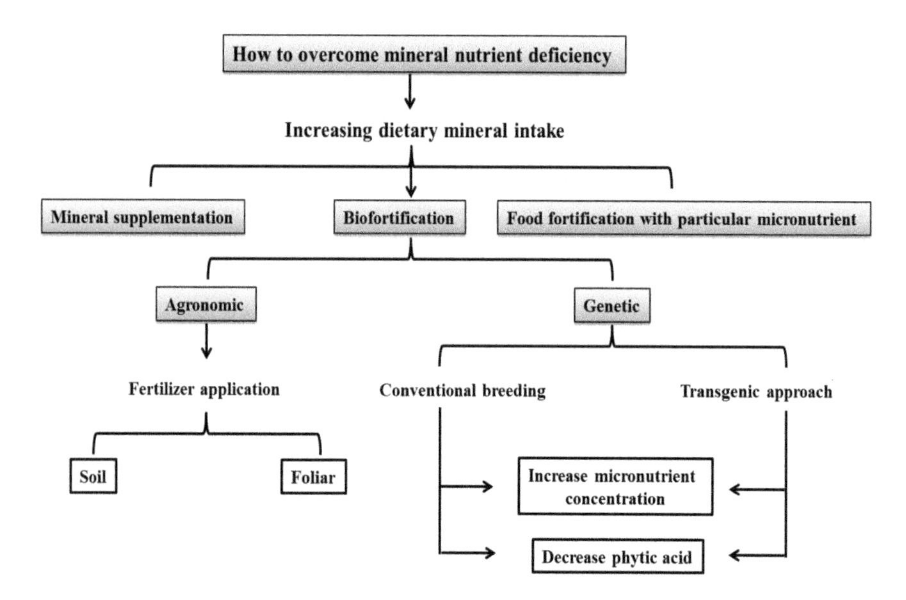

FIGURE 6.3 Diagrammatic representation of various biofortification strategies to remediate the hidden hunger crisis.

TRANSCRIPTOMICS IN BUCKWHEAT CROP IMPROVEMENT

Transcriptomics is one of the robust functional genomics approach for characterization of the candidate genes that are involved in the regulation of diverse biological processes (Kumar et al. 2016). This technique helps to provide exhaustive information regarding the expression patterns of genes and functional polymorphism, particularly in orphan crops with poorly characterized genomes, and such information could be an important source for functional marker development like single nucleotide polymorphism (SNPs), intron-spanning region (ISR) and expressed sequence tags-simple sequence repeats (EST-SSRs). Previous reports revealed that comprehensive transcriptomics analysis in buckwheat has already been analyzed for floral structure, salt tolerance and aluminum toxicity (Logacheva et al. 2011; Wu et al. 2017; Xu et al. 2017; Chen et al. 2017). Furthermore, in Tartary buckwheat, about 11,676 differentially expressed genes have been characterized using a transcriptomics approach, particularly during the seed developmental stage, and such candidate genes act as important genomic resources for functional characterization pertaining to various traits using a reverse genetics approach (Huang et al. 2017). Further, Illumina HiSeq 2000 was carried out in sweet buckwheat for whole-genome sequencing that resulted in assembly of the genome in 387,594 scaffolds, constituting about 1.17 Gb of data. A total of 35,816 annotated genes have been identified through gene prediction analysis and used to generate a Buckwheat Genome Database (BGDB; http://buckwheat.kazusa.or.jp) (Yasui et al. 2016). Further, this genome sequence information possesses an important application in identifying the genes that are involved in the biosynthesis of 2S-albumin type allergens, flavonoids and granule-bound starch synthases and in regulating heteromorphic self-incompatibility (Matsui and Yasui 2020).

PROTEOMICS AND METABOLOMICS IN BUCKWHEAT CROP IMPROVEMENT

Proteomics is another advanced approach that provides an insight into plant genomes via proteome mapping, protein–protein interactions and comparative proteomics as reported by Varshney et al. (2009). Further, buckwheat proteomics unravels the novel proteins, helps to better understand the various developmental changes and provides in-depth analysis via functional genomics (Lee et al. 2016). However, in buckwheat, the proteomic approach to understanding the traits associated with grain quality has yet to achieve results, although some studies pertaining to proteomic analysis have been reported that facilitate an understanding of the light-dependent developmental physiology of buckwheat (Lee et al. 2016). Another bottleneck in the buckwheat improvement program is the presence of some allergic proteins that act as hindrance for its commercialization (Heffler et al. 2014). Therefore, for identifying such allergic proteins in buckwheat, absolute quantitative proteomics is a potential approach to this challenge. To date, the non-availability of a well-assembled reference genome sequence is the major lacuna in the large-scale genome-wide proteomic study of buckwheat. The recently available genome sequences of buckwheat species help to promote a better understanding of seed proteomics (Yasui et al. 2016; Wang et al. 2017; Zhang et al. 2017). Besides, proteomics is an important tool for biologists and breeders in identifying the novel gene (Collard and Mackill 2008).

In addition to proteomics, metabolomics is another omics tool in identifying and characterizing the novel metabolites that contribute to the nutritional profile of buckwheat. From this perspective, it is necessary to unravel the important metabolites that find wide application in the functional food sector for human consumption. To date, high-throughput metabolomics studies in buckwheat have not been reported, although some data pertaining to metabolomics analysis of Tartary buckwheat have been reported by some authors (Thwe et al. 2013; Kumar et al. 2016; Pirzadah et al. 2017). Thwe et al. (2013) reported 47 metabolites using gas chromatography–time-of-flight mass spectrometry (GC-TOFMS) that are involved in the phenylpropanoid pathway. For a better understanding of the gene regulation mechanism, multi-omics could be an interdisciplinary approach that integrates metabolomics data with other omics data. Further, for the buckwheat biofortification breeding program, it is necessary to investigate the complete seed metabolome and integrate it with other omics data on seed tissues, such as transcriptomics and proteomics data.

TISSUE CULTURE–BASED APPROACH TO BUCKWHEAT CROP IMPROVEMENT

A tissue culture–based approach is another important step in crop improvement programs (Murashige and Skoog 1977). In previous studies, the development of calluses with the ability for organogenesis and the restoration of plantlets from the isolated cotyledons and hypocotyls of buckwheat has been reported (Yamane 1974). Furthermore, Woo et al. (2000a, 2000b) carried out research on *F. esculentum* for the development of a plant regeneration system that could have widespread genetic transformation applications. Somatic embryogenesis is another tissue-culture approach that has been carried out in *F. esculentum* and *F. tataricum* by culturing immature embryos, as reported by various authors (Neskovic et al. 1987; Rumyantseva et al. 1989;Lachmann and Adachi 1990; Rumyantseva et al. 1989). Woo et al. (1999a) reported an effective protocol for the isolation of viable protoplasts from egg cells in buckwheat that could be fused with somatic protoplasts and other gametoplasts due to their biological function. Transformation via *in planta* approach has also revolutionized the transgenic technology by escaping the somaclonal variations and protracted tissue culture cycle (Kojima et al. 2000a, 2000b). Kwon et al. (2013) reported an effective method for plant regeneration in *F. esculentum* using hypocotyl segments as explants. Similarly, a hairy root culture under *in vitro* conditions via *A. rhizogenes* mediated transformation to enhance flavonoid content has been reported in Tartary buckwheat (Park et al. 2011; Gabr et al. 2012; Thwe et al. 2013).

ARTIFICIAL MUTATIONS IN THE SELF-COMPATIBLE LINES

Self-compatible buckwheat lines have been developed that offer many advantages in buckwheat breeding programs over self-incompatible lines, viz. easy development of segregating lines for gene analysis via self-pollination, easy fixation of morphological and genotypes and seed development without insect pollinators, often leading to the production of stable yields minimally affected by various abiotic/biotic stresses (Matsui et al. 2008; Matsui and Yasui 2020). As there is a high level of genetic diversity among the landraces of *F. esculentum*, therefore various essential recessive traits are predicted to be suppressed by dominant traits when in heterozygous state (Ohnishi 1993; Mizuno and Yasui 2019). However, crossing between self-compatible lines and landraces results in the formation of self-compatible progeny, in which masked beneficial traits may emerge. Further, the self-compatible lines harboring a mutation induced by various mutagens like gamma ray and ethyl methanesulfonate are convenient tools for forward genetics. Therefore, the self-compatible lines are efficient at generating novel qualitative traits. Moreover, another powerful tool known as "targeting induced local lesion in genomes" (TILLING) is used to identify a mutated gene. The development of "waxy wheat" (in which the endosperm of the wheat is devoid of amylose) is one of the breakthroughs of this technology as reported by Slade et al. (2005, 2012). This important tool could be used to identify beneficial alleles like non-allergenic/glutinous starch particularly from landraces and mutation-induced populations of buckwheat. In addition, buckwheat is also altered by various abiotic factors, viz. photoperiod and temperature stress. Therefore, the production of self-compatible buckwheat lines that are adapted to a particular region would be a cumbersome process and thus needs to be addressed. In this regard, we propose an effective breeding protocol that integrates both self-compatible and self-incompatible buckwheat varieties and would be very effective in generating high-value commercial varieties in a particular region (Figure 6.4). The self-compatible varieties could also be used to generate lines that possess beneficial traits, viz. being low amylose/non-allergenic, by homogenization of recessive dysfunctional genes. Identification of a beneficial trait is followed by the introduction of this trait into the leading self-incompatible cultivar via various rounds of backcrossing, and it is expected that the newly developed variety would possess the similar desired traits because of its similar gene construct and genetic makeup; thus the newly formed self-incompatible elite variety possessing the beneficial traits will be achieved (Figure 6.4). While selecting new self-incompatible elite varieties, novel self-compatible varieties may also be established. However, the backcrossing of the beneficial recessive gene (dysfunctional) and S^h allele would lead to the formation of self-compatible buckwheat lines with desired and beneficial traits, and in this protocol, molecular marker-associated selection is an efficient means to trace the S^h allele. After various rounds of selfing, if the self-compatible buckwheat line does not exhibit any indication of inbreeding depression, it can be regarded as a novel self-compatible variety with valuable traits and thus exhibiting stable high-yield production (Matsui and Yasui 2020).

GENETIC ENGINEERING FOR BUCKWHEAT CROP IMPROVEMENT

The incorporation of novel candidate genes in buckwheat using a genetic engineering approach is another tool to combat the inherent problems associated with buckwheat. Various transformation events have been well established in buckwheat where *Agrobacterium tumefaciens* and *A. rhizogenes* are utilized as potential vectors (Table 6.5) (Park et al. 2011; Gabr et al. 2012). Currently, it has been reported that the *TrMYB4* transcription factor regulates rutin biosynthesis in the hairy root culture of *F. cymosum* induced by *A. rhizogenes* and, in the near future, could be a versatile tool that utilizes buckwheat metabolic engineering (Luo et al. 2020). Recently, CRISPR/Cas9 technology, a genome editing tool, has revolutionized the functional genomics field by addressing productivity constraints, particularly in major crops (Bortesi and Fischer 2015; Ma et al. 2016). However, the application of CRISPR/Cas9 technology has not been implemented in buckwheat crop improvement. The precise nucleotide modifications in genes encoding rutinosidase using a genome-editing tool has the potential to unravel the genetic mechanism of rutinosidase activity in *F. tataricum* that is responsible for its bitter taste (Suzuki and Morishita 2016). Moreover, the development of rutin-rich non-bitter functional food products of Tartary buckwheat would be possible via CRISPR/Cas9-based targeted molecular stacking of rutinosidase-inhibiting genes. This genome editing technology also finds great application in the knockdown of genes that encode allergic proteins in the seed proteome of *F. esculentum*, thus making it more promising and attractive in the functional food sector.

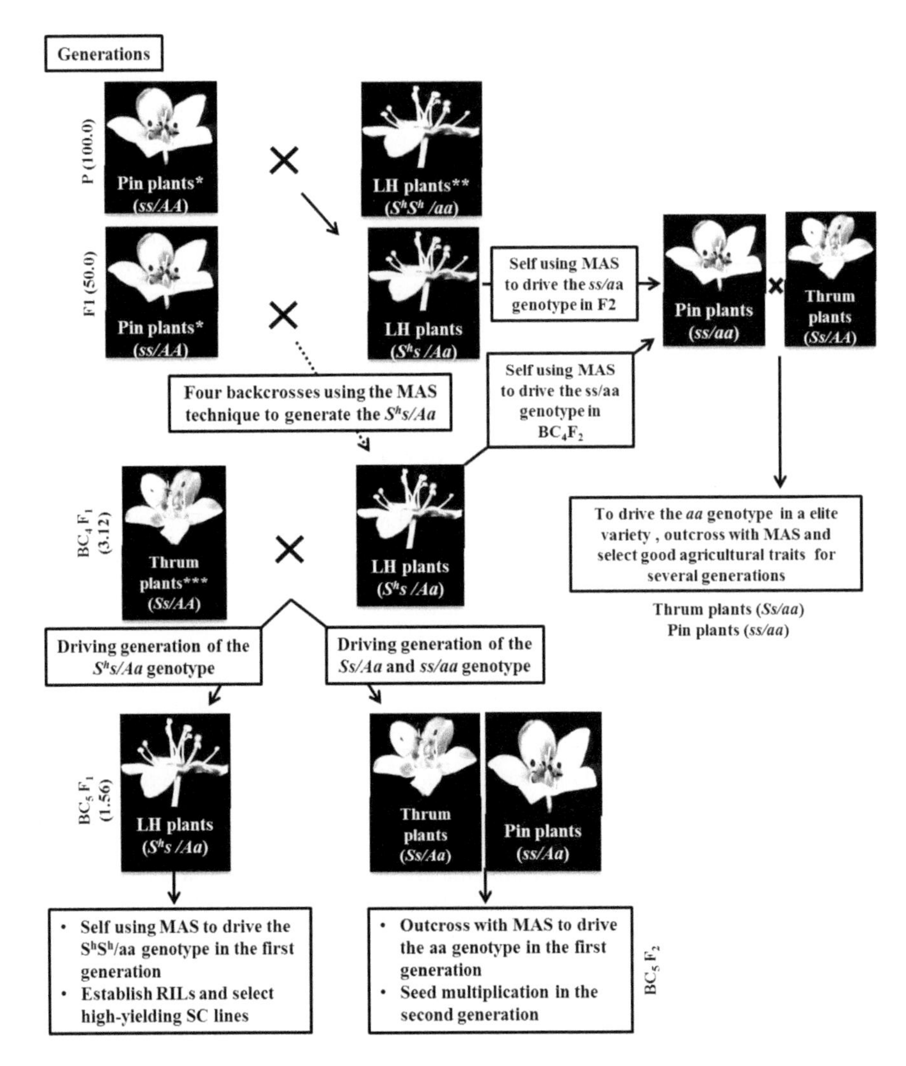

FIGURE 6.4 Schematic diagram of a buckwheat breeding system using SI and SC plants. The values in parentheses indicate the percentage of genetic contribution from a parental line. *S*, allele controlling the thrum flower phenotype; *s*, allele controlling the pin flower phenotype; S^h, allele controlling the long-homostyle flower phenotype. Dominance relationship among the alleles: $S > S^h > s$. A, Dominant allele of the wild type; a, dysfunctional valuable allele; *, pin plant from the SI elite variety; **, SC long-homostyle (LH) plant obtained from mutagenesis and/or crosses between SI landraces and SC experimental plants, such as "Norin-PL1." ***, thrum plants from the SI elite variety. In the BC_4F_1 generation, many thrum plants should be used to avoid inbreeding depression in later generations. MAS, marker-assisted selection.

TABLE 6.5
A Comprehensive List of Various Transformation Events in Buckwheat

Buckwheat species	Method of transformation	Explant used	Outcome	References
F. esculentum	*Agrobacterium rhizogenes*-mediated	Stem	Transgenic hairy root clones producing high quantity of rutin	Kim et al. (2010)
F. tataricum	*A. rhizogenes*-mediated	Stem	Transgenic hairy root cultures producing high amount of caffeic acid, chlorogenic acid and rutin	Park et al. (2011)
F. esculentum	*A. rhizogenes*-mediated	Excised roots, leaves and stems	Transgenic hairy cultures producing higher quantity of chlorogenic, *p*-hydroxybenzoic, *p*-anisic and caffeic acids	Gabr et al. (2012)
F. esculentum	*A. tumefaciens*-mediated	Cotyledon	Establishment of transformation efficiency	Miljus-Djukic et al. (1992)
F. esculentum	*A. tumefaciens*-mediated	Hypocotyls	Establishment of transformation efficiency	Kim et al. (2001)
F. esculentum	*in planta*	Apical meristem	Establishment of efficient *in planta* transformation protocol	Kojima et al. (2000a, 2000b)

CONCLUSION AND FUTURE PERSPECTIVE

As buckwheat possesses a high nutritional profile and has immense potential in the functional food sector, it is regarded as a golden crop to tackle the future invisible hunger crisis. To date, buckwheat has not received much attention, particularly from biotechnologists, in terms of remediating the inherent problems associated with this crop. Moreover, the available genetic variations of buckwheat enriched with a high nutritional profile have not made any major impact on the development of high-yielding cultivars. Thus, the development of a database to unravel the genetic potential of buckwheat germplasm is the need of the hour. While Joshi et al. (2020) have reported some discoveries pertaining to buckwheat genomics, these are not at par with other crops. Currently, few data regarding intervention in buckwheat genomics have been reported, but in the near future, the field of buckwheat genomics could be revolutionized using the latest technologies. Wild species of buckwheat could act as an important source of genes/alleles that could be employed in various buckwheat breeding programs. Moreover, multi/interdisciplinary research that involves multi-omics approaches, viz. genomics, transcriptomics, proteomics, ionomics and metabolomics, is needed to promote this neglected crop. A graphic representation of a proposal involving the use of integrative genomics and breeding approach for accelerating buckwheat crop improvement is illustrated in Figure 6.5. Extensive research pertaining to

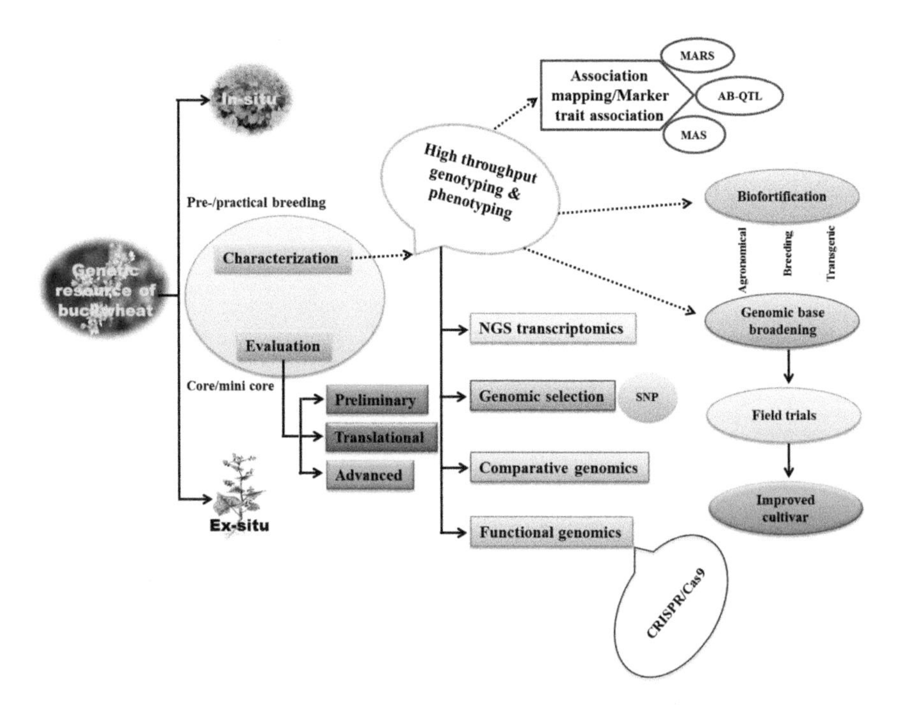

FIGURE 6.5 Graphic representation involving integrative genomics and breeding approach for accelerating buckwheat crop improvement.

buckwheat value addition and biofortification in conjugation with molecular pharming is another emerging concept to combat the future hidden hunger crisis. Lastly, research efforts are needed using a collaborative approach, particularly on the part of agriculturists, biotechnologists, agronomist, food scientists and other stakeholders, to promote and accelerate buckwheat crop improvement; moreover, public awareness and government action plans for marketing could undoubtedly make buckwheat the golden crop of the future.

REFERENCES

Adhikari, K. N. and Campbell, C. G. 1998. *In vitro* germination and viability of buckwheat (*Fagopyrum esculentum* Moench) pollen. *Euphytica* 102(1): 87–92.

Ali, J., Nagano, M., Penner, G., Campbell, C. and Adachi, T. 1998. Identification of RAPD markers linked to homostylar (Ho) gene in buckwheat [*Fagopyrum*]. *Japan J Breed* 48(1): 59–62.

Allen, S. and de Brauw, A. 2018. Nutrition sensitive value chains: Theory, progress, and open questions. *Glob Food Secur* 16: 22–28.

Amano, T., Zhu, Q., Wang, Y., Inoue, N. and Tanaka, H. 1993. Case studies on high yields of paddy rice in Jiangsu Province, China. II. Analysis of characters related to lodging. *Jpn J Crop Sci* 62(2): 275–281.

Asaduzzaman, M., Minami, M., Matsushima, K. and Nemoto, K. 2009. An *in-vitro* ovule culture technique for producing interspecific hybrid between Tartary buckwheat and common buckwheat. *J Biol Sci* 9(1): 1–11.

Bjorkman, T. 1995. Gametophyte selection through pollen competition in buckwheat. *Curr Adv Buckwheat Res* 443–451.

Bohanec, B., Neskovic, M. and Vujicic, R. 1993. Anther culture and androgenetic plant regeneration in buckwheat (*Fagopyrum esculentum* Moench). *Plant Cell Tissue Org* 35(3): 259–266.

Bortesi, L. and Fischer, R. 2015. The CRISPR/Cas9 system for plant genome editing and beyond. *Biotechnol Adv* 33(1): 41–52.

Campbell, C. G. 1995. Inter-specific hybridization in the genus *Fagopyrum*. In: *Proceedings of the 6th International Symposium on Buckwheat*, pp. 255–263.

Campbell, C. G. 1997 *Buckwheat. Fagopyrum esculentum Moench. Promoting the Conservation and Use of Underutilized and Neglected Crops. 19.* IPK, Germany and IPGRI, Rome, Italy.

Chauhan, R. S., Gupta, N., Sharma, S. K., Rana, J. C., Sharma, T. R. and Jana, S. 2010. Genetic and genome resources in buckwheat – present and future perspectives. *Eur J Plant Sci Biotechnol* 4: 33–44.

Chen, Q. F., Hsam, S. L. K. and Zeller, F. J. 2007. Cytogenetic studies on diploid and autotetraploid common buckwheat and their autotriploid and trisomics. *Crop Sci* 47(6): 2340–2345.

Chen, Q. F., Huang, X. Y., Li, H. Y., Cui, Y. S. and Cui, Y. 2018. Recent progress in perennial buckwheat development. *Sustainability* 10(2): 536.

Chen, W. W., Xu, J. M., Jin, J. F., Lou, H. Q., Fan, W. and Yang, J. L. 2017. Genome wide transcriptome analysis reveals conserved and distinct molecular mechanisms of Al resistance in buckwheat (*Fagopyrum esculentum* Moench) leaves. *Int J Mol Sci* 18(9): 1859.

Collard, B. C. Y. and Mackill, D. J. 2008. Marker-assisted selection: An approach for precision plant breeding in the twenty-first century. *Philos Trans R Soc Lond B* 363(1491): 557–572.

Dar, F. A., Pirzadah, T. B., Malik, B., Tahir, I. and Rehman, R. U. 2018. Molecular genetics of buckwheat and its role in crop improvement. In: Zhou, M., Kreft, I., Tang, Y. and Suvorova, G. (eds.) *Buckwheat Germplasm in the World*, 1st edn. Elsevier Publications, USA, pp. 271–286.

FAO 2017. *The Future of Food and Agriculture: Trends and Challenges.* Food & Agriculture Organization of the United Nations, Rome.

FAOSTAT 2018. Production-yield quantities of buckwheat in world + (total) 1961–2016. http://www.fao.org/faost at/en/#data/QC/visua lize. Accessed 19 July 2018.

Fesenko, A. N., Fesenko, N. N., Romanova, O. I. and Fesenko, I. N. 2016. Crop evolution of buckwheat in eastern Europe: Micro evolutionary trends in the secondary center of buckwheat genetic diversity. In: Zhou, M., Kreft, I., Woo, S. H., Chrungoo, N. and Wieslander, G. (eds.) *Molecular Breeding and Nutritional Aspects of Buckwheat.* Academic Press, Cambridge, pp. 99–108.

Fesenko, I. N. and Fesenko, N. N. 2010. New species form of buckwheat – *Fagopyrum hybridum. Vestn Orel GAU* 4: 78–81.

Fesenko, I. N., Fesenko, N. N. and Onishi, O. 2001. Compatibility and congruity of interspecific crosses in *Fagopyrum.* In: *Proceedings of the 8th International Symposium on Buckwheat.* Korea, pp. 404–410.

Fesenko, N. V., Fesenko, N. N., Romanova, O. I., Alekseeva, E. C. and Suvorova, G. N. 2006. *Theoretical Basis of Plant Breeding, vol 5. The Gene Bank and Breeding of Groat Crops: Buckwheat.* VIR, St. Petersburg.

Gabr, A., Sytar, O., Ahmed, A. and Smetanska, I. 2012. Production of phenolic acid and antioxidant activity in transformed hairy root cultures of common buckwheat (*Fagopyrum esculentum* M). *Aust J Basic Appl Sci* 6: 577–586.

Gupta, N., Sharma, S. K., Rana, J. C. and Chauhan, R. S. 2012. AFLP fingerprinting of Tartary buckwheat accessions (*Fagopyrum tataricum*) displaying rutin content variation. *Fitoterapia* 83(6): 1131–1137.

Hara, T., Iwata, H., Okuno, K., Matsui, K. and Ohsawa, R. 2011. QTL analysis of photoperiod sensitivity in common buckwheat by using markers for expressed sequence tags and photoperiod-sensitivity candidate genes. *Breed Sci* 61(4): 394–404.

Heffler, E., Pizzimenti, S., Badiu, I., Guida, G. and Rolla, G. 2014. Buckwheat allergy: An emerging clinical problem in Europe. *J Allergy Ther* 5: 2.

Hirose, T., Lee, B. S., Okuno, J., Konishi, A., Minami, M. and Ujihara, A. 1995a. Interspecific pollen–pistil interaction and hybridization in genus *Fagopyrum.* In: *Proceedings of the 6th International Symposium on Buckwheat Japan*, pp. 239–245.

Hirose, T., Ujihara, A., Kitabayashi, H. and Minami, M. 1995b. Pollen tube behavior related to self-incompatibility in interspecific crosses of *Fagopyrum. Breed Sci* 45(1): 65–70.

Huang, J., Deng, J., Shi, T., Chen, Q., Liang, C., Meng, Z., Zhu, L., Wang, Y., Zhao, F., Yu, S. and Chen, Q. 2017. Global transcriptome analysis and identification of genes involved in nutrients accumulation during seed development of rice Tartary buckwheat (*Fagopyrum tataricum*). *Sci Rep* 7(1): 11792.

Huang, X., Han, B. 2014. Natural Variations and Genome-Wide Association Studies in Crop Plants. *Ann Rev Plant Biol.* 65(1): 531–551.

Joshi, D. C., Chaudhari, G. V., Sood, S., Kant, L., Pattanayak, A., Zhang, K., Fan, Y., Janovska, D., Meglic, V. and Zhou, M. 2019. Revisiting the versatile buckwheat: Reinvigorating genetic gains through integrated breeding and genomics approach. *Planta* 250(3): 783–801.

Joshi, D. C., Zhang, K., Wang, C., Chandora, R., Khurshid, M., Li, J., He, M., Georgiev, M. I. and Zhou, M. 2020. Strategic enhancement of genetic gain for nutraceutical development in buckwheat: A genomics-driven perspective. *Biotechnol Adv* 39: 107479.

Kim, H., Kang, H., Lee, Y., Lee, S., Ko, J. and Rha, E. 2001. Direct regeneration of transgenic buckwheat from hypocotyl segment by agrobacterium-mediated transformation. *Kor J Crop Sci* 46: 375–379.

Kim, H., Kim, J. K., Kang, L., Jeong, K. and Jung, S. 2010. Docking and scoring of quercetin and quercetin glycosides against alpha-amylase receptor. *Bull Korean Chem Soc* 31(2): 461–463.

Kim, Y., Kim, S., Lee, K., Chang, K., Kim, N., Shin, Y. and Park, C. 2002. Interspecific hybridization between Korean buckwheat landraces (*Fagopyrum esculentum* Moench) and self-fertilizing buckwheat species (*F. homotropicum* Ohnishi). *Fagopyrum* 19: 37–42.

Kojima, M., Arai, Y., Iwase, N., Shirotori, K., Shiori, H. and Nozue, M. 2000a. Development of a simple and efficient method for transformation of buckwheat plants (*Fagopyrum esculentum*) using *Agrobacterium tumefaciens. Biosci Biotechnol Biochem* 64(4): 845–847.

Kojima, M., Hihahara, M., Shiori, H., Nozue, M., Yamomoto, K., Sasaki, T. and Sasaki, T. 2000b. Buckwheat transformed with cDNA of a rice MADS box gene is stimulated in branching. *Plant Biotechnol* 17(1): 35–42.

Konishi, T. and Ohnishi, O. 2006. A linkage map for common buckwheat based on microsatellite and AFLP markers. *Fagopyrum* 23: 1–6.

Korte, A. and Farlow, A. 2013. The advantages and limitations of trait analysis with GWAS: A review. *Plant Methods* 9: 29.

Kreft, M. 2016. Buckwheat phenolic metabolites in health and disease. *Nutr Res Rev* 29(1): 30–39.

Krotov, A. S. and Golubeva, E. A. 1973. Cytological studies on an interspecific hybrid *Fagopyrum tataricum* × *F. cymosum. Bull Appl Bot Genet Plant Breed* 51: 256–260.

Kumar, A., Metwal, M., Kaur, S., Gupta, A. K., Puranik, S., Singh, S., Singh, M., Gupta, S., Babu, B. K., Sood, S. and Yadav, R. 2016. Nutraceutical value of finger millet [*Eleusine coracana* (L.) Gaertn.], and their improvement using omics approaches. *Front Plant Sci* 7: 1–14.

Kwon, S. J., Han, M. H., Huh, Y. S., Roy, S. K., Lee, C. W. and Woo, S. H. 2013. Plantlet regeneration via somatic embryogenesis from hypocotyls of common buckwheat (*Fagopyrum esculentum* Moench). *Korean J Crop Sci* 58(4): 331–335.

Lachmann, S. and Adachi, T. 1990. Callus regeneration from hypocotyl protoplasts of Tartary buckwheat (*Fagopyrum tataricum* Gaertn). *Fagopyrum* 10: 62–64.

Lee, D. G., Woo, S. H. and Choi, J. S. 2016. Biochemical properties of common and Tartary buckwheat: Centered with buckwheat proteomics. In: Zhou, M., Kreft, I., Woo, S. H., Chrungoo, N. and Wieslander, G. (eds.) *Molecular Breeding and Nutritional Aspects of Buckwheat.* Academic Press, Cambridge, pp. 239–259.

Liu, M., Zheng, T., Ma, Z., Wang, D., Wang, T., Sun, R., He, Z., Peng, J. and Chen, H. 2016. The complete chloroplast genome sequence of Tartary buckwheat cultivar Miqiao (*Fagopyrum tataricum* Gaertn.). *Mitochondrial DNA B* 1(1): 577–578.

Logacheva, M. D., Kasianov, A. S., Vinogradov, D. V., Samigullin, T. H., Gelfand, M. S., Makeev, V. J. and Penin, A. A. 2011. De novo sequencing and characterization of floral transcriptome in two species of buckwheat (Fagopyrum). *BMC Genom* 12: 30.

Luo, Q., Li, J., Wang, C., Cheng, C., Shao, J., Hui, J., Zeng, Y., Wang, J., Zhu, X. and Xu, Y. 2020. *TrMYB4* transcription factor regulates the rutin biosynthesis in hairy roots of *F. cymosum. Plant Sci* 294: 110440.

Ma, K. H., Kim, N. S., Lee, G. A., Lee, S. Y., Lee, J. K., Yi, J. Y., Park, Y. J., Kim, T. S., Gwag, J. G. and Kwon, S. J. 2009. Development of SSR markers for studies of diversity in the genus *Fagopyrum. Theor Appl Genet* 119(7): 1247–1254.

Ma, X., Zhu, Q., Chen, Y., Liu, Y. G. and Y-g, L. 2016. CRISPR/Cas9 platforms for genome editing in plants: Developments and applications. *Mol Plant* 9(7): 961–974.

Marshall, H. 1969. Isolation of self-fertile, homomorphic forms in buckwheat *Fagopyrum sagittatum* Gilib. *Crop Sci* 9(5): 651–653.

Matsui, K. and Yasui, Y. 2020. Buckwheat heteromorphic self-incompatibility: Genetics, genomics and application to breeding. *Breed Sci* 70(1): 32–38.

Matsui, K., Kiryu, Y., Komatsuda, T., Kurauchi, N., Ohtani, T. and Tetsuka, T. 2004. Identification of AFLP makers linked to non-seed shattering locus (Sht1) in buckwheat and conversion to STS markers for marker-assisted selection. *Genome* 47(3): 469–474.

Matsui, K., Tetsuka, T., Hara, T. and Morishita, T. 2008. Breeding and characterization of a new self-compatible common buckwheat parental line "Buckwheat Norin-PL1". *Res Bull Natl Agric Res Cent Kyushu Okinawa Reg* 49: 11–17.

Miljus-Djukic, J., Neskovic, M., Ninkovic, S. and Crkvenjakov, R. 1992. *Agrobacterium* mediated transformation and plant regeneration of buckwheat (*Fagopyrum esculentum* Moench). *Plant Cell Tiss Org Cult* 29(2): 101–108.

Mizuno, N. and Yasui, Y. 2019. Gene flow signature in the S-allele region of cultivated buckwheat. *BMC Plant Biol* 19(1): 125.

Morishita, T., Hara, T. and Hara, T. 2020. Important agronomic characteristics of yielding ability in common buckwheat; ecotype and ecological differentiation, preharvest sprouting resistance, shattering resistance, and lodging resistance. *Breed Sci* 70(1): 39–47.

Murai, M. and Ohnishi, O. 1996. Population genetics of cultivated common buckwheat, *Fagopyrum esculentum* Moench. X. Diffusion routes revealed by RAPD markers. *Genes Genet Syst* 71(4): 211–218.

Murashige, T. and Skoog, F. 1977. Manipulation of organ initiation in plant tissue cultures. *Bot Bull Acad Sin* 18: 1–24.

Nagano, M., Aii, J., Kuroda, M., Campbell, C. and Adachi, T. 2001. Conversion of AFLP markers linked to the Sh allele at the S locus in buckwheat to a simple PCR based marker form. *Plant Biotechnol* 18(3): 191–196.

Neskovic, M., Vujicic, R. and Budimir, S. 1987. Somatic embryogenesis and bud formation from immature embryos of buckwheat (*Fagopyrum esculentum* Moench). *Plant Cell Rep* 6(6): 423–426.

Neskovic, M., Culafic, L. and Vujicic, R. 1995. Somatic embryogenesis in buckwheat (*Fagopyrum* Mill.) and sorrel (*Rumex L.*), *Polygonaceae*. In: Bajaj, Y. P. S. (ed.) *Biotechnology in Agriculture and Forestry, vol 31. Somatic Embryogenesis and Synthetic Seed II*. Springer-Veriag, Berlin Heidelberg, pp. 412–427.

Niroula, R. K., Bimb, H. P. and Sah, B. P. 2006. Interspecific hybrids of buckwheat (*Fagopyrum* spp.) regenerated through embryo rescue. *Sci World* 4: 74–77.

Ohnishi, O. 1993. Population genetics of cultivated common buckwheat *Fagopyrum esculentum* Moench. VIII. Local differentiation of land races in Europe and the silk road. *Jpn J Genet* 68(4): 303–316.

Pan, S. J. and Chen, Q. F. 2010. Genetic mapping of common buckwheat using DNA, protein and morphological markers. *Hereditas* 147(1): 27–33.

Park, N., Li, O., Uddin, R. and Park, S. 2011. Phenolic compound production by different morphological phenotypes in hairy root cultures of *Fagopyrum tataricum* Gaertn. *Arch Biol Sci* 63(1): 193–198.

Pirzadah, T. B., Malik, B., Tahir, I., Qureshi, M. I. and Rehman, R. U. 2017. Metabolite fingerprinting and antioxidant potential of Tartary Buckwheat – an underutilized pseudocereal crop from Kashmir Region. *Free Radic Antioxid* 7(1): 95–106.

Pirzadah, T. B., Malik, B., Tahir, I., Hakeem, K. R. and Rehman, R. U. 2020. Lead toxicity alters the activities of antioxidant enzymes and modulate the biomarkers in Tartary buckwheat plants. *Int Biodeterior Biodegrad* 151, http://www.ncbi.nlm.nih.gov/pubmed/104992.

Przybylski, R. and Gruczynska, E. 2009. A review of nutritional and nutraceutical components of buckwheat. *Eur J Plant Sci Biotechnol* 3: 10–22.

Quinet, M., Cawoy, V., Lefevre, I., Van-Miegroet, F., Jacquemart, A. L. and Kinet, J. M. 2004. Inflorescence structure and control of flowering time and duration by light in buckwheat (*Fagopyrum esculentum* Moench). *J Exp Bot* 55(402): 1509–1517.

Rana, J. C. and Sharma, B. D. 2000. Variation, genetic divergence and interrelationship analysis in buckwheat. *Fagopyrum* 17: 9–14.

Rana, J. C., Singh, M., Chauhan, R. S., Chahota, R. K., Sharma, T. R., Yadav, R. and Archak, S. 2016. Genetic resources of buckwheat in India. In: Zhou, M., Kreft, I., Woo, S. H., Chrungoo, N. and Wieslander, G. (eds.) *Molecular Breeding and Nutritional Aspects of Buckwheat*. Academic Press, Cambridge, pp. 109–135.

Rauf, M., Yoon, H., Lee, S., Hyun, D. Y., Lee, M. C., Oh, S. and Choi, Y. M. 2020. Evaluation of *Fagopyrum esculentum* Moench germplasm based on agro-morphological traits and the rutin and quercetin content of seeds under spring cultivation. *Genet Resour Crop Evol* 67(6): 1385–1403.

Rumyantseva, N., Sergeena, N., Khakinova, L., Salnikov, V., Gumerova, E. and Lozovaya, V. 1989. Organogenesis and somatic embroyogenesis in tissue culture of two buckwheat species. *Fiziol Rast* 36: 187–194.

Rumyantseva, N., Fedoseeva, N., Abdrakhmanova, G., Nikolskaya, V. and Lopato, S. 1995. Interspecific hybridization in the genus *Fagopyrum* using in vitro embryo culture. In: *Proceedings of the 6th International Symposium on Buckwheat*. Japan, pp. 211–220.

Samimy, C., Bjorkman, T., Siritunga, D. and Blanchard, L. 1996. Overcoming the barrier to interspecific hybridization of *Fagopyrum esculentum* with wild *Fagopyrum tataricum*. *Euphytica* 91(3): 323–330.

Shaikh, N. Y., Guan, L. M. and Adachi, T. 2002. Ultrastructural aspects on degeneration of embryo, endosperm and suspensor cells following interspecific crosses in genus Fagopyrum. *Breed Sci* 52(3): 171–176.

Sharma, T. and Jana, S. 2002. Species relationships in *Fagopyrum* revealed by PCR-based DNA fingerprinting. *Theor Appl Genet* 105(2–3): 306–312.

Shin, D. H., Kamal, A., Yun, Y. H., Bae, J. S., Lee, Y. S., Lee, M. S., Chung, K. Y. and Woo, S. H. 2009. Enhanced seed development in the progeny from the interspecific backcross (*Fagopyrum esculentum*; *F. homotropicum*) *F. esculentum*. *Korean J Plant Resour* 22: 209–214.

Slade, A. J., Fuerstenberg, S. I., Loeffler, D., Steine, M. N. and Facciotti, D. 2005. A reverse genetic, nontransgenic approach to wheat crop improvement by TILLING. *Nat Biotechnol* 23(1): 75–81.

Slade, A. J., McGuire, C., Loeffler, D., Mullenberg, J., Skinner, W., Fazio, G., Holm, A., Brandt, K. M., Steine, M. N., Goodstal, J. F. and Knauf, V. C. 2012. Development of high amylose wheat through TILLING. *BMC Plant Biol* 12: 69.

Small, E. 2017. Buckwheat – the world's most biodiversity-friendly crop? *Biodiversity* 18(2–3): 108–123.

Song, J. Y., Lee, G. A., Yoon, M. S., Ma, K. H., Choi, Y. M., Lee, J. R., Jung, Y., Park, H. J., Kim, C. K. and Lee, M. C. 2011. Analysis of genetic diversity and population structure of buckwheat (*Fagopyrum esculentum* Moench) landraces of Korea using SSR markers. *Korean J Plant Resour* 24(6): 702–711.

Song, Y., Fang, Q., Jarvis, D., Bai, K., Liu, D., Feng, J. and Long, C. 2019. Network analysis of seed flow, a traditional method for conserving Tartary buckwheat (*Fagopyrum tataricum*) landraces in Liangshan, Southwest China. *Sustainability* 11(16): 4263.

Suvorova, G. and Zhou, M. 2018. Distribution of cultivated buckwheat resources in the world. In: Zhou, M., Kreft, I., Tang, Y. and Suvorova, G. (eds.) *Buckwheat Germplasm in the World*, 1st edn. Elsevier Publications, USA, , pp. 21–35.

Suvorova, G. N. 2001. The problem of interspecific cross of *Fagopyrum esculentum* Moench. × *Fagopyrum cymosum* Meissn. In: *Proceedings of the 8th International Symposium on Buckwheat*. Korea, pp. 311–318.

Suvorova, G. N. 2010. Perspectives of interspecific buckwheat hybridization. In: *Proceedings of the 11th International Symposium on Buckwheat*. Russia, pp. 295–299.

Suvorova, G. N., Fesenko, N. N. and Kosturbin, M. M. 1994. Obtaining of interspecific buckwheat hybrid (*Fagopyrum esculentum* Moench, *Fagopyrum cymosum* Meissn.). *Fagopyrum* 14: 13–16.

Suzuki, T. and Morishita, T. 2016. Bitterness generation, rutin hydrolysis and development of trace rutinosidase variety in Tartary buckwheat. In: Zhou, M., Kreft, I., Woo, S. H., Chrungoo, N. and Wieslander, G. (eds.) *Molecular Breeding and Nutritional Aspects of Buckwheat*. Academic Press, Cambridge, pp. 239–259.

Tang, Y., Sun, J. X., Peng, D. C., Liu, J. L. and Shao, J. R. 2011. Study on the nutrients and medical elements in wild buckwheat. *J Sichuan Higher Inst Cuisine* 6: 28–31.

Thwe, A. A., Kim, J. K., Li, X., Kim, Y. B., Uddin, M. R., Kim, S. J., Suzuki, T., Park, N. I. and Park, S. U. 2013. Metabolomic analysis and phenylpropanoid biosynthesis in hairy root culture of Tartary buckwheat cultivars. *PLOS ONE* 8(6):e65349.

Tsuji, K. and Ohnishi, O. 2000. Origin of cultivated Tartary buckwheat (*Fagopyrum tataricum* Gaertn) revealed by RAPD analyses. *Genet Resour Crop Evol* 47(4): 431–438.

Ujihara, A., Nakamura, Y. and Minami, M. 1990. Interspecific hybridization in genus Fagopyrum-properties of hybrids (*F. esculentum* Moench, *F. cymosum* Meissner) through ovule culture. In: *Gamma Field Radiation Breeding*. NIAR, MAFF, Japan, pp. 45–51.

Varshney, R. K., Nayak, S. N., May, G. D. and Jackson, S. A. 2009. Next-generation sequencing technologies and their implications for crop genetics and breeding. *Trends Biotechnol* 27(9): 522–530.

Wang, C. L., Ding, M. Q., Zou, C. Y., Zhu, X. M., Tang, Y., Zhou, M. L. and Shao, J. R. 2017. Comparative analysis of four buckwheat species based on morphology and complete chloroplast genome sequences. *Sci Rep* 7(1): 6514.

Wang, Y., Scarth, R. and Campbell, C. 2002. Interspecific hybridization between *Fagopyrum tataricum* (L) Gartn and *F. esculentum* Moench. *Fagopyrum* 19: 31–35.

Watson, A., Ghosh, S., Williams, M. J., Cuddy, W. S., Simmonds, J., Rey, M., Hatta, M. A. M., Hinchliffe, A., Steed, A., Reynolds, D. et al. 2018. Speed breeding is a powerful tool to accelerate crop research and breeding. *Nat Plants* 4(1): 23–29.

Woo, S. H., Tsai, Q. and Adachi, T. 1995. Possibility of interspecific hybridization by embryo rescue in the genus *Fagopyrum*. *Curr Adv Buckwheat Res* 6: 225–237.

Woo, S. H., Adachi, T. and Park, S. I. 1998. Breeding of a new autogamous buckwheat: 2 Seed protein analysis and identification of RAPD markers linked to the Ho (Sh) gene. *Korean J Plant Resour* 30: 144–145.

Woo, S. H., Adachi, T., Jong, S. K. and Campbell, C. G. 1999a. Isolation of protoplasts from viable egg cells of common buckwheat (*Fagopyrum esculentum* Moench). *Can J Plant Sci* 79(4): 593–595.

Woo, S. H., Wang, Y. J. and Campbell, C. 1999b. Interspecific hybrids with *Fagopyrum cymosum* in the genus *Fagopyrum*. *Fagopyrum* 16: 13–18.

Woo, S. H., Adachi, T., Jong, S. K. and Campbell, C. G. 2000a. Isolation of protoplasts from viable sperm cells of common buckwheat (*Fagopyrum esculentum* Moench). *Can J Plant Sci* 80(3): 583–585.

Woo, S. H., Nair, A., Adachi, T. and Campbell, C. G. 2000b. Plant regeneration from cotyledon tissues of common buckwheat (*Fagopyrum esculentum* Moench). *Vitro Cell Dev* 36(5): 358–361.

Woo, S. H., Campbell, C. G. and Jong, S. K. 2002. Interspecific buckwheat hybrid between *Fagopyrum esculentum* and *F. cymosum* through embryo rescue. *Korean J Breed* 34: 322–327.

Woo, S. H., Kim, S. H., Tsai, K. S., Chung, K. Y., Jong, S. K., Adachi, T. and Choi, J. S. 2008. Pollen-tube behavior and embryo development in interspecific crosses among the genus *Fagopyrum*. *J Plant Biol* 51(4): 302–310.

Woo, S. H., Kamal, A. H. M., Tatsuro, S., Campbell, C. G., Adachi, T., Yun, S. H., Chung, K. Y. and Choi, J. S. 2010. Buckwheat (*Fagopyrum esculentum* Moench.): Concepts, prospects and potential. *Eur J Plant Sci Biotech* 4: 1–16.

Wu, Q., Bai, X., Zhao, W., Xiang, D., Wan, Y., Yan, J., Zou, L. and Zhao, G. 2017. De novo assembly and analysis of Tartary buckwheat (*Fagopyrum tataricum* Garetn) transcriptome discloses key regulators involved in salt-stress response. *Genes* 8(10): 255.

Xiaolei, D., Zongwen, Z., Bin, W., Yanqin, L. and Anhu, W. 2013. Construction and analysis of genetic linkage map in Tartary buckwheat (*Fagopyrum tataricum*) using SSR. *Chin Agric Sci Bull* 29: 61–65.

Xu, J. M., Fan, W., Jin, J. F., Lou, H. Q., Chen, W. W., Yang, J. L. and Zheng, S. J. 2017. Transcriptome analysis of Al-induced genes in buckwheat (*Fagopyrum esculentum* Moench) root apex: New insight into Al toxicity and resistance mechanisms in an Al accumulating species. *Front Plant Sci* 8: 1141.

Yabe, S., Hara, T., Ueno, M., Enoki, H., Kimura, T., Nishimura, S., Yasui, Y., Ohsawa, R. and Iwata, H. 2014. Rapid genotyping with DNA micro-arrays for high-density linkage mapping and QTL mapping in common buckwheat (*Fagopyrum esculentum*). *Breed Sci* 64(4): 291–299.

Yabe, S., Hara, T., Ueno, M., Enoki, H., Kimura, T., Nishimura, S., Yasui, Y., Ohsawa, R. and Iwata, H. 2018. Potential of genomic selection in mass selection breeding of an allogamous crop: An empirical study to increase yield of common buckwheat. *Front Plant Sci* 9: 276.

Yamane, Y. 1974. Induced differentiation of buckwheat plants from subcultured calluses *in vitro*. *Japan J Genet* 49(3): 139–146.

Yang, W., Hao, Y., Li, G. and Zhou, N. 1998. Relationship between reproductive growth of common buckwheat and light duration. In: *Proceedings of the 7th International Symposium on Buckwheat*. Winnipeg, Manitoba, Canada, pp. 44–48.

Yasui, Y., Wang, Y., Ohnishi, O. and Campbell, C. G. 2004. Amplified fragment length polymorphism linkage analysis of common buckwheat (*Fagopyrum esculentum*) and its wild self-pollinated relative *Fagopyrum homotropicum*. *Genome* 47(2): 345–351.

Yasui, Y., Hirakawa, H., Ueno, M., Matsui, K., Katsube-Tanaka, T., Yang, S. J., Aii, J., Sato, S. and Mori, M. 2016. Assembly of the draft genome of buckwheat and its applications in identifying agronomically useful genes. *DNA Res* 23(3): 215–224.

Yu, J. H., Kwon, S. J., Choi, J. Y., Ju, Y. H., Roy, S. K., Lee, D. G., Park, C. H. and Woo, S. H. 2019. Variation of rutin and quercetin in Tartary buckwheat germplasm. *Fagopyrum* 36: 51–65.

Zhang, L., Li, X., Ma, B., Gao, Q., Du, H., Han, Y., Li, Y., Cao, Y., Qi, M., Zhu, Y. et al. 2017. The Tartary buckwheat genome provides insights into rutin biosynthesis and abiotic stress tolerance. *Mol Plant* 10(9): 1224–1237.
Zhao, G., Tang, Y. and Wang, A. H. 2002. Research on the nutrient constituents and medicinal values of *Fagopyrum cymosum* seeds. *Chin Wild Plant Resour* 21: 39–41.
Zheng, S., Cheng-hua, H. A. N. and Huang, K. F. 2011. Research on Se content of different Tartary buckwheat genotypes. *Agric Sci Technol* 12: 102–104.
Zhou, M., Kreft, I., Woo, S. H., Chrungoo, N. and Wieslander, G. 2016 *Molecular Breeding and Nutritional Aspects of Buckwheat.* Academic Press, Cambridge, p. 482.

Index